Experimental Zoology

THE AUTHORS

Prof. (Dr.) Ramesh Mahadeo Gejage [M. Sc., Ph. D., F. S. E. Sc. (Dumka), FAZ (Pune)] is presently working as Associate Professor in Zoology. He obtained his M. Sc. and Ph. D. degree in Zoology from S. U., Kolhapur, (M.S.) in the year 1994 and 2005 respectively. He entered in the teaching profession as Assistant Professor in Zoology at, Smt. K. R. P. Kanya Mahavidyalaya, Islampur Tal. Walwa, Dist. Sangli (M.S.) in the year 1994. He has worked as N.S.S. programme officer from 1999-2003. He worked as adviser Kanya Nature Club in the year 2004 affiliated with WWF nature India, Kolhapur Division. He organized three days "National Symposium on Recent Trends in Life Sciences" on Feb., 2007 at College. He has been awarded nine awards and recently awarded as "Adarsh Shikshak Puraskar" on September, 2018 from Giants Welfare Foundation, Islampur. His major research interests are in the fields of environmental biology, physiology and enzymology. He has published 136 research articles out of this 31 are in peer reviewed and reputed scientific International, National and State level research journals. He has attended 77 seminar/conferences out of this 6 are International and 45 are National level. Under his guidance 2 research scholars have been awarded for Ph. D. and 02 awarded for M. Phil degree. Currently, he is guiding 02 Ph. D. scholars. He has completed UGC minor research project. He participated in study group of Holar samaj organized by BARTI, Pune. He published two books "Chordate Animals" and "Human Reproductive Physiology" in the year 2021. He is managing editor for *Journal of Association of Zoologists,* Pune. He is member, BOS in Zoology V.G. Vaze College of Arts, Science and Commerce (Autonomous), Mulund, Mumbai from 2020. He is member of BOS subcommittee in Zoology, S.U. Kolhapur from 2018. He is life member of ISCAP (Tirupati) from 2002; *Indian Journal of Environment and Ecoplanning* (Dumka); *Indian Journal of Environmental Protection* (Varanasi); *ZOOPASTA*, Kolhapur and *AZI*, Pune.

Dr. (Mrs.) Manisha Ramesh Gejage [M. Sc., M. Phil., Ph. D.] is presently working as Assistant Professor in Zoology. She obtained her M.Sc., M. Phil. and Ph. D. degree in Zoology from Shivaji University, Kolhapur, Maharashtra in the year 2006, 2009 and 2013 respectively. She entered in the teaching profession as Assistant Professor in Zoology at, VNAC and BNS Mahavidyalaya, Shirala Tal. Shirala, Dist. Shirala (M.S.) in the year 2009. She awarded "Samajbhusan Award" from Harivittal Foundation, Kadlas, Tal. Sangola, Dist. Solapur. Her major research interests are in the fields of biodiversity of insects and entomology. She has published 81 research articles out of this 21 are in peer reviewed and reputed scientific International, National and State level research journals. She has attended 42 seminar/conferences out of this 6 are International and 28 are National level. She has published three books "Taxonomy of Crickets and Household Pests" on 2009; "Human Reproductive Physiology" and "Chordate Animals".

Experimental Zoology

— *Authors* —

Prof. (Dr.) Ramesh M. Gejage

M. Sc., Ph.D., DIT, F. S. E. Sc. (Dumka), FAZI (Pune)
Associate Professor and Head, Department of Zoology,
Smt. Kusumtai Rajarambapu Patil Kanya Mahavidyalaya
(Arts, Commerce and Science), Islampur Tal. Walwa,
Dist. Sangli-415 409 (M. S.) India.

Dr. (Mrs.) Manisha R. Gejage

M. Sc., M. Phil., Ph. D., MHCET
Assistant Professor in Zoology,
Department of Zoology,
V. N. A. C. and B. N. Science Mahvidyalaya,
Shirala Tal. Shirala, Dist. Sangli-415 408 (M. S.) India.

2023

Daya Publishing House®
A Division of
Astral International Pvt. Ltd.
New Delhi – 110 002

ISBN: 9789359191263

Published by : **Daya Publishing House®**
A Division of
Astral International Pvt. Ltd.
– ISO 9001:2015 Certified Company –
4736/23, Ansari Road, Darya Ganj
New Delhi-110 002
Ph. 011-43549197, 8130496929
E-mail: info@astralint.com
Website: www.astralint.com

Preface

An *experiment* is a procedure carried out to support a hypothesis. The *scientific method* is a process for experimentation that is used to explore observations and answer questions. While writing this book we have put together all our combined resources and have tried to simplify the subject as much as practicable. We have taken topics from syllabus of practicals suggested by (CBCS) Indian Universities for undergraduate program in Zoology. In this book, We have given information on animal specimens with reference to morphological peculiarities and classification upto orders from chordates. Characters identifying venomous and non-venomous snakes with suitable examples. Studies on common birds. Brain of fowl and hyoid apparatus, sclerotic plates, pecten, collumella of fowl. Temporary preparations of cycloid, ctenoid and placoid scales in fishes. Qualitative tests to identify functional groups of carbohydrates and lipid. The activity of salivary amylase under optimum conditions including effect of temperature, pH and salinity on activity of salivary amylase. Estimation of total protein and lipid, DNA isolation and estimation of uric acid. Economic importance of insects. Study of arthropod vectors associated with human diseases. Estrus cycle of rat including histology of reproductive organs. Surgical techniques like ovarectomy, hysterectomy, castration and vasectomy in rat. Study of animal house. Human vaginal exfoliate cytology. Study of sperm count and sperm motility. Contraceptive devices. How to write educational tour report and laboratory preparations.

The authors are grateful to Principal Hon. R. D. Sawant Secretary, K. E. Society and Principal, Dr. Vilas Kale, Smt. K. R. P. Kanya Mahavidyalaya, Islampur, Principal, Dr. Rajendra Bansode, V.N.A.C. and B.N. Science Mahvidyalaya, Shirala. Principal Dr. S. B. Kengar, Prof. Pradeep Patil, Dr. Mrs. Sujata Patil, Dr. Raghunath Kengar, Dr. J. J.

Pol, Prof. Dr. V.Y. Deshapande, Prof. Dr. R. D. Bodare, Mr. Baban Javir, Adv. Kiran Javir, Mr. Sunil Bhajanawale, Prof. Dr. Bira Parse, Dr. Ajit Kengar, Dr. Yuvaraj Kengar, Mr. Raghunath Awate, Mrs. Ushalata Awate, Mr. Harshad Awate, Mr. Baburao Gejage, Mr. Bhagavan Gejage, Mr. Shivaji Gejage, Mr. Namdev Gejage, mother Parvati, our daughter Rasika and son Vipul for their encouragement.

Ramesh M. Gejage

Manisha R. Gejage

Contents

Chapter 1

Animal Diversity

1. STUDY OF CHORDATES

Aim: Morphological Peculiarities and Classification up to the Orders.

1.1. *Herdmania*

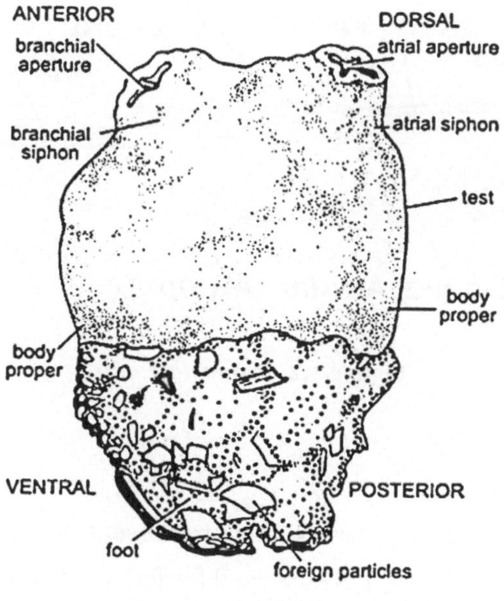

Figure 1.1: ***Herdmania.***

Classification

Phylum : Chordata - Nerve cord, notochord and gill slits are present.

Subphylum : Urochordata- Notochord present in larval tail.

Class : Ascidiacea- Larvae well developed than adult.Gill slits many.

Order : Pleurogona- Body undivided, Neural gland dorsal.

Type : *Herdmania*

Morphological Peculiarities of *Herdmania*

1. It is marine water animal. It suddenly contracts body and water is given out through branchial and atrial aperture hence called sea squirt.
2. It is hermaphrodite animal (monoecious). Fertilization is external, indirect development and larval stage called tadpole larva. Pipette shaped spicules are present in the tunic. They are sessile, oval and broad base attached to substratum.
3. It feeds on algae, diatoms, zooplanktons and phytoplanktons.
4. The basal foot is attached with sand particles, shell pieces and other foreign particles. Body is enclosed in a soft leathery body wall called test and animals from this group is commonly called tunicates. Large pharynx with gill slits.

1.2. *Branchiostoma*

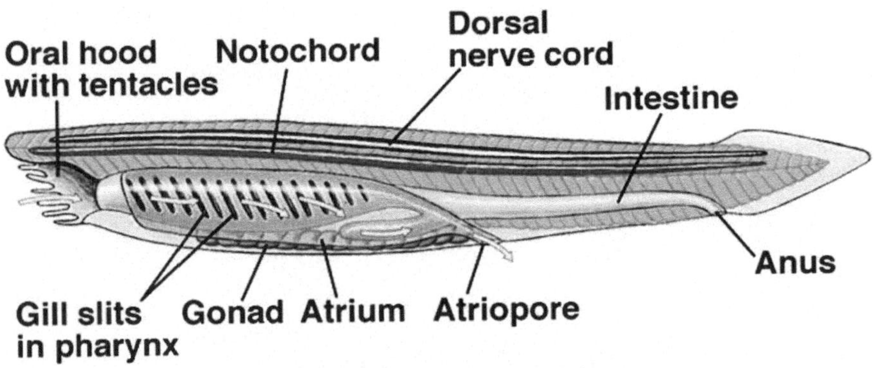

Figure 1.2: *Branchiostoma*.

Classification

Phylum : Chordata

Subphylum : Cephalochordata-Notochord present through entire body length.

Class : Leptocardii- Many gill slits used for feeding and not for respiration.

Order : Amphioxiformes- Body pointed at both the ends.

Type : *Branchiostoma*

Morphological Peculiarities of *Branchiostoma*

1. It is found in shallow, sandy, brackish water and pointed at both the ends hence also called Amphioxus or lancelet.
2. Body is fish like laterally compressed and pointed at both the ends.
3. Body divisible into trunk and tail. Distinct head absent. Single layered epidermis. Respiration through body wall. Anus present at the base of the tail.
4. Tapered anterior part of trunk is called rostrum.
5. The expanded oral hood gives oral cirri (10-12) around the mouth.
6. Atriopore is present just behind the symmetrical metapleural fold. Dorsal, ventral and caudal fin present. Fins are continuous with one another. Dorsal and ventral fin have fin ray boxes.
7. Persistent type of notochord present throughout the length of body. Myotomes (muscle bundles) are 'V' shaped and present on either side of the body (62 pairs). Sexual dimorphism absent. Gonads many (27 pairs).

1.3. *Petromyzon*

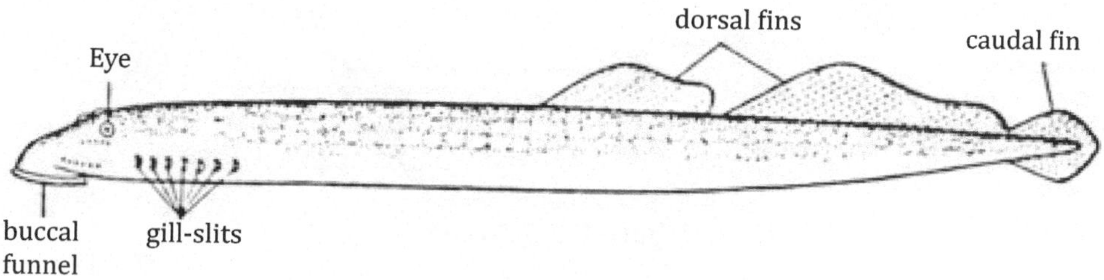

Figure 1.3: *Petromyzon.*

Classification

Phylum : Chordata

Subphylum : Vertebrata- Vertebral column is present.

Superclass : Agnatha- Paired appendages absent. Upper and lower jaws absent.

Class : Cyclostomata- Circular mouth present.

Order : Petromyzontia or Petromyzontiformes, Ammocoetes larva present.

Type : *Petromyzon*

Morphological Peculiarities of *Petromyzon*

1. It is marine water and freshwater animal. It is also called lamprey.
2. Body eel like, cylindrical and distinguished in to head, trunk and tail. Skin soft and glandular. Upper lip well developed and forms buccal funnel.

3. Head and trunk is cylindrical and tail is laterally compressed.
4. Mouth circular and suctorial provided with suckers and rows of horny teeth. Tongue raping. Eyes well developed without eyelids.
5. Nasal orifice at the top of the head and ends blindly. Seven pairs of external gill slits present near the head. Mouth without tentacles.
6. Two dorsal fins and a caudal fin present. Fin supported by fin rays.
7. Small sensory pores of lateral line system extend along the lateral side of the body below the head. It is ecto-parasitic on other fishes.
8. It migrates from sea to river for spawning (anadromous animal). Fresh water eel fish migrates from river to marine water called as catadromous fish.

1.4. *Sphyrna*

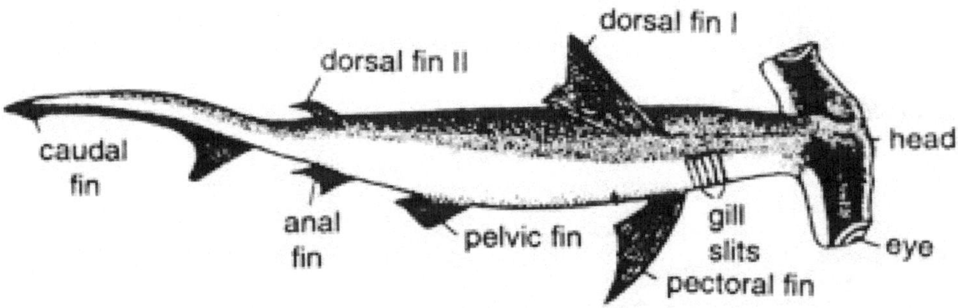

Figure 1.4: *Sphyrna*.

Classification

Phylum : Chordata

Subphylum : Vertebrata

Superclass : Pisces-All fishes. Cold blooded fishes.

Class : Chondrichthyes- Operculum absent. Cartilagenous fishes.

Subclass : Selachi or (Elasmobranchii)-Mouth ventral, placoid scales present.

Superorder : Pleurotremata - Five pairs of gills. Air bladder absent.

Order : Carcharhiniformes- Head with two lobes (Hammer like head).

Type : *Sphyrna*

Morphological Peculiarities of *Sphyrna*

1. *Sphyrna* commonly called hammer headed shark. Head has two lateral lobes that resemble a hammer. The body divisible into head, trunk and tail.

2. Two prominent eyes one on each lobe of the head. Nictitating membrane present.
3. Mouth ventral with upper and lower jaws. Nostril ventral in position.
4. Placoid scales are present on the body.
5. Five pairs of external gill slits present just behind the each eye. Paired pectoral and pelvic fins present and unpaired dorsal, ventral (anal fin) and caudal fins present. First dorsal fin is larger than second dorsal fin. Caudal fin with an upper precaudal pit. Homodont type of teeth present in both jaws.
6. The two lobes of fin are unequal externally as well as internally and is turned upwards called as heterocercal tail. Male *Sphyrna* fish has a pair of copulatory organ called claspers.

1.5. *Pristis*

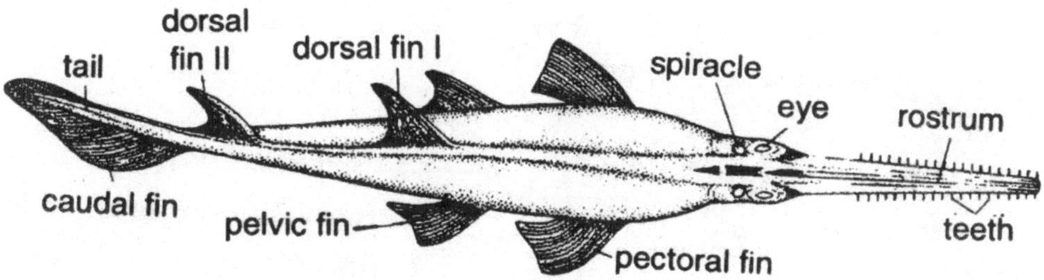

Figure 1.5: *Pristis*.

Classification

Phylum : Chordata

Subphylum : Vertebrata

Superclass : Pisces

Class : Chondrichthyes

Subclass : Selachi or Elasmobranchii

Superorder : Hypotremata- Ventral many gill slits present.

Order : Rajiformes - No electrical organ, long flattened rostrum present.

Type : *Pristis*

Morphological Peculiarities of *Pristis*

1. It is commonly called sawfish or carpenter shark.
2. It is characterized by a long, narrow, flattened rostrum, or nose extension.
3. Rostrum is lined with sharp transverse teeth which are arranged in a way that resembles a saw.

4. Body is divisible into head, trunk and tail.
5. Anterior part dorso-ventrally flattened like ray fish and posterior part is shark-like.
6. It exhibits close relationship with rays.
7. Skin is smooth and without scales.
8. Head contains a pair of eyes.
9. A pair of spiracles present behind the eyes.
10. It is viviparous.

1.6. *Torpedo*

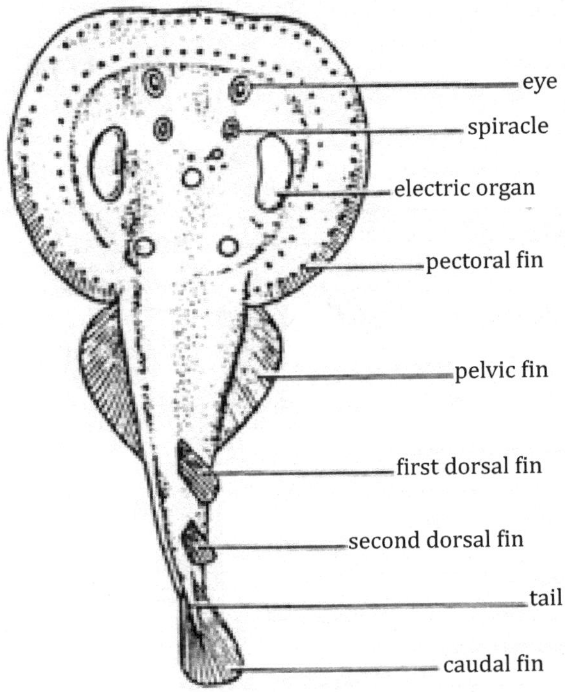

Figure 1.6: *Torpedo*

Classification

Phylum : Chordata

Subphylum : Vertebrata

Superclass : Pisces

Class : Chondrichthyes

Subclass : Selachi or Elasmobranchii

Superorder : Hypotremata- Ventral many gill slits present.

Order : Torpediniformes- Electric organs present.

Type : *Torpedo*

Morphological Peculiarities of *Torpedo*

1. It is also known electric ray because of presence of a pair of electric organs. Electric organs present in between head and pectoral fins. Body is divisible into anterior semicircular disk and posterior tail. Nasofrontal lobe quadrangular. Eyes located just above the electric organs. Spiracles present behind the eyes.
2. It is viviparous fish. Tail is thick with two dorsal and one caudal fin.
3. Skin is smooth and without scales. Five pairs of gill slits located ventrally.
4. Eyes are located near the electric organs.
5. Pectoral fins fused with the head and body. First and second dorsal fins and a caudal fin present.

1.7. *Labeo*

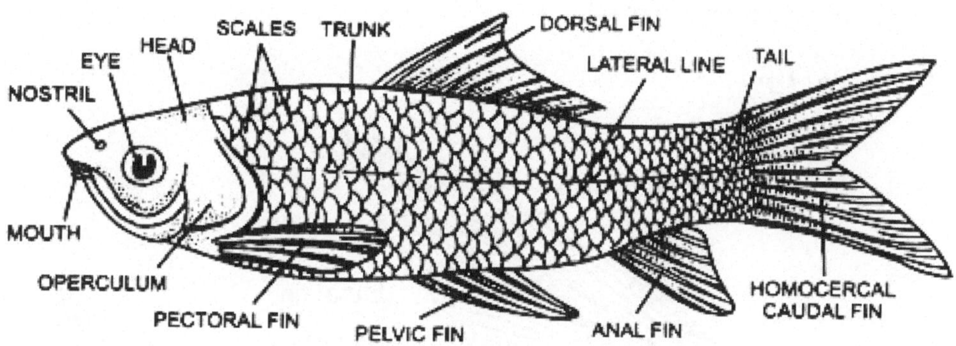

Figure 1.7: *Labeo.*

Classification

Phylum : Chordata

Subphylum : Vertebrata

Superclass : Pisces

Class : Osteichthyes-Bony fishes.

Subclass : Teleostomi or Actinopterygii- Modern fishes, Nostril does not communicate with into mouth, cycloid scales present.

Superorder : Teleostei- Four pairs of filamentous gills. Ganoine layer absent in scale.

Paired fins without a central axis.

Order : Cypriniformes- Body oblong and compressed. Pelvic and anal fins are abdominal. Gut connected to swim bladder by a duct. Head scaleless. Pharyngeal teeth present.

Type : *Labeo*

Morphological Peculiarities of *Labeo*

1. It is commonly found in the fresh water ponds, lakes and river. It is found in middle layer of pond. Body is divisible in to head, trunk and tail.
2. Head is depressed. Mouth inferior with thick lips for collection of food from middle layer of water body. Paired nostrils and paired eyes present. Barbels help in detection of food. Jaw without teeth.
3. Pharangeal teeth present. Dorsal, anal, caudal, paired pectoral and anal fin with soft fin rays present. Caudal fin is homocercal with equal lobes. Dorsal fin ahead of pelvic fins. Caudal fin forked for direction of movement.
4. Anal fin short with 7 or 8 rays for perfect balancing. Lateral line complete. Cycloid scales are present on the skin. Dorsal, pectoral, pelvic, ventral and caudal fins are present. Complete lateral line present. Lips thick and fleshy. Homocercal tail.

1.8. *Exocoetus*

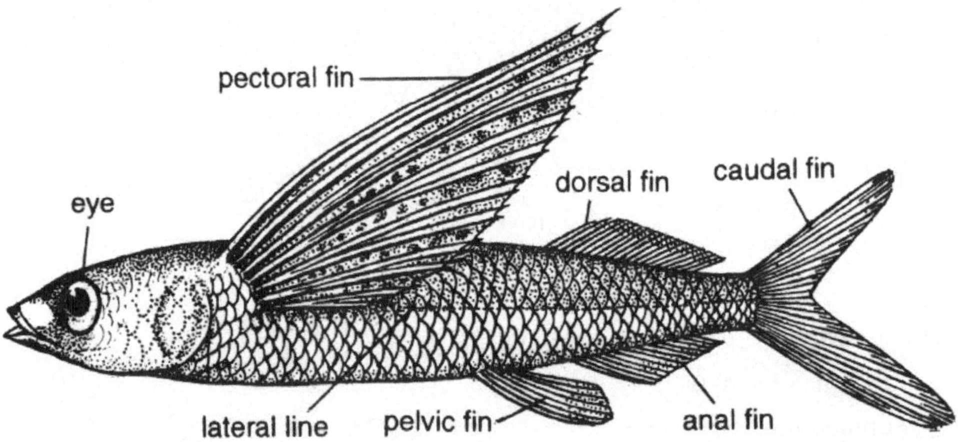

Figure 1.8 : *Exocoetus*.

Classification

Phylum : Chordata

Subphylum : Vertebrata

Superclass : Pisces

Class : Osteichthyes

Subclass : Teleostomi or Actinopterygii

Superorder : Teleostei- Four pairs of gills. Ganoine layer absent in scale.

Paired fins without a central axis.

Order : Beloniformes or Synentognathi- Pectoral fins large and shifted dorsally. Body elongated, spindle shaped, flying fishes. Cycloid scales present,

Type : *Exocoetus*

Morphological Peculiarities of *Exocoetus*

1. It is also called flying fish. It makes gliding flights. Elongated body divisible into head, trunk and tail. Ventral fin used for swimming.
2. Body covered with overlapping cycloid scales. Spindle shaped body minimizes water resistance during swimming. Pectoral fins with black patches.
3. Mouth small with teeth in both jaws. Lower lip larger than upper for collection of food from upper surface. Snout short and blunt. Homocercal forked tail fin.
4. Dorsal and ventral (anal) fins are used in balancing and propelling. They are supported by fin rays. Pectoral fins are large, wing like and shifted dorsally. It can glide over the surface of the water for about 400 meters.

1.9. *Anguilla*

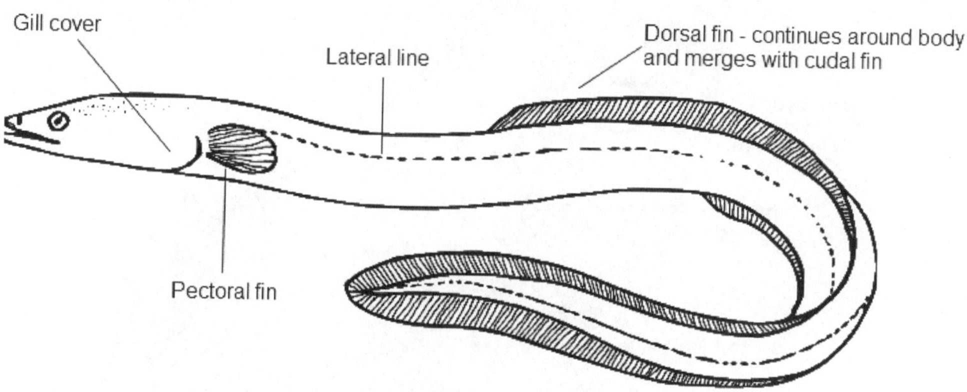

Figure 1.9: *Anguilla.*

Classification

Phylum : Chordata

Subphylum : Vertebrata

Superclass : Pisces

Class : Osteichthyes

Subclass : Teleostomi or Actinopterygii

Order : Anguiliformes or apodes: Dorsal, ventral and caudal fins continued.

Type : *Anguilla*

Morphological Peculiarities of *Anguilla*

1. It is called fresh water eel fish. It can live several hours outside the water.
2. Body is slender elongated and snake like. It is divisible into head, trunk and tail.
3. Head bears mouth, eyes and nostril. Four pairs of gills are covered by operculum.
4. It migrates from river to marine water (Catadromous fish).
5. Dorsal, anal (ventral) and caudal fins are joined together forming continuous fin.
6. Pelvic fin absent. Small pectoral fins present behind the operculum.
7. In fresh water eel scales are very small and so deeply embedded that the fish appears to be naked.

1.10. *Bufo*

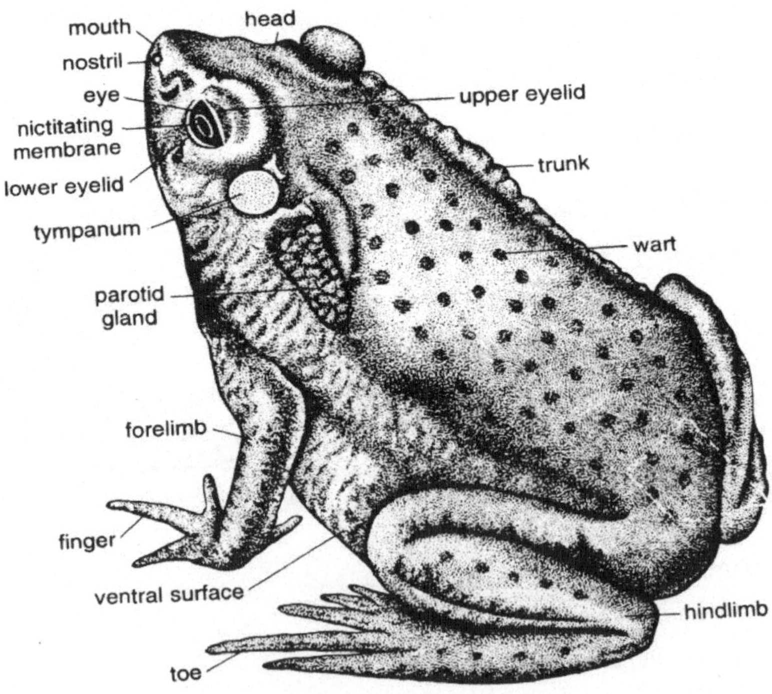

Figure 1.10: *Bufo*.

Classification

Phylum : Chordata

Subphylum : Vertebrata

Superclass : Tetrapoda- A pair of fore limbs and hind limbs present.

Class : Amphibia-Cold blooded. Aquatic and terrestrial in habitat.

Order : Anura or salientia- Tail and gills present in larval stage only.

Type : *Bufo*

Morphological Peculiarities of *Bufo*

1. It is called common toad. It is terrestrial, nocturnal, hiding under stones or in burrows in day. It enters water only to breed and spawn.
2. It has rough, dry and warty skin with poison and mucous glands. It moves with a slow walk or short jumps. Exoskeleton is absent.
3. Vocal sac absent. Body divisible into head and trunk. Head bears mouth, large eyes, nostril and tympanum. A pair of poisonous glands (paratoid) present just behind the eyes. Toes of the fore feet are turned inwards.
4. At breeding time, the male develops nuptial pads on the first three fingers.
5. Female is brown and male is grey in color. The life of common toad is about ten to twelve years.

1.11. *Hyla*

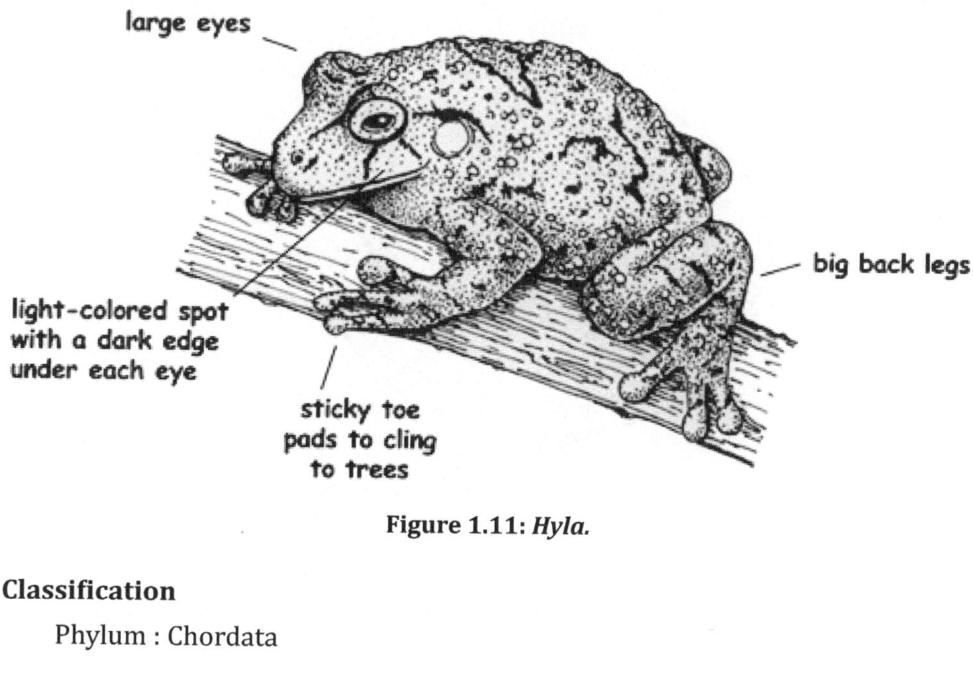

Figure 1.11: ***Hyla.***

Classification

Phylum : Chordata

Subphylum : Vertebrata

Superclass : Tetrapoda

Class : Amphibia

Order : Anura or salientia

Type : *Hyla*

Morphological Peculiarities of *Hyla*

1. It is called tree frog. It is arboreal. It moves with a slow walk on the tree. Body divisible into head and trunk. Head bears mouth, large eyes, nostril and tympanum. Skin green colored dorsally and yellowish-green ventrally.
2. A distinct neck absent. Fore limbs and hind limbs are short. Tree frogs typically have well-developed discs at the finger and toe tips.
3. Skin smooth and hygroscopic.
4. Teeth present upper in jaws.
5. Large bulging eyes present.
6. Fingers and toes are webbed. Tongue is free at behind.
7. Male with expanded vocal sacs.
8. The fingers, toes and limbs are small.
9. Toes long and provided with adhesive discs.

1.12. *Ichthyophis*

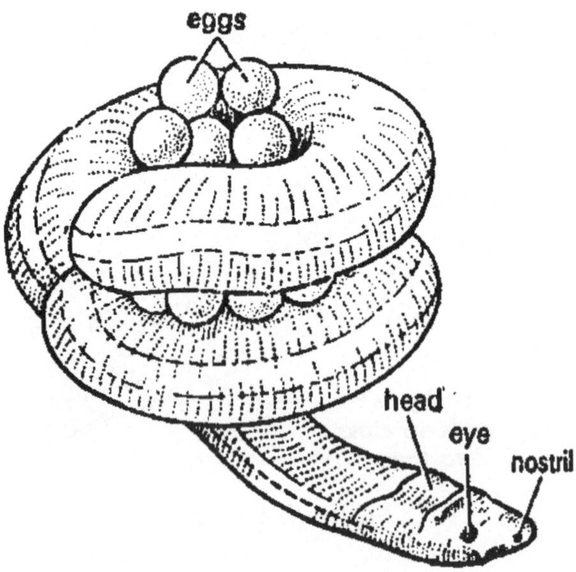

Figure 1.12: *Ichthyophis.*

Classification

Phylum : Chordata

Subphylum : Vertebrata

Superclass : Tetrapoda

Class : Amphibia

Order : Apoda or Gymnophiona: Fore limbs and hind limbs are absent.

Type : *Ichthyophis*

Morphological Peculiarities of *Ichthyophis*

1. The body is worm like slender divided into head, trunk and tail.
2. Head bears eyes, nostrils and a pair of sensory tentacles.
3. Eyes small, functionless and covered by skin. Left lung is reduced.
4. Limbs, girdles and sternum absent. Short tail and anus is sub-terminal.
5. Parental care shows body coiling around the eggs. Long trachea and right lung.
6. The tentacles are used for tasting food. Middle ear and tympanum are absent.
7. It is a worm-like caecilian ring-like groove on the body.
8. Head is small, depressed, microscopic blind eyes and protrusible tentacle.

1.13. *Salamandra*

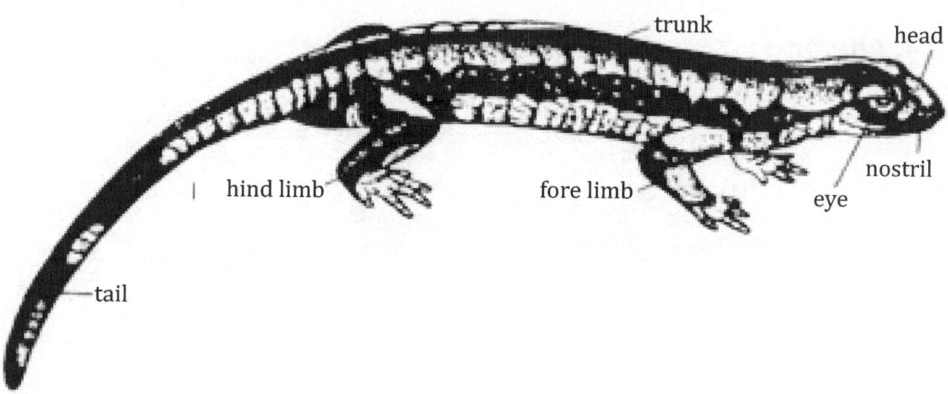

Figure 1.13: ***Salamandra.***

Classification

Phylum : Chordata

Subphylum : Vertebrata

Superclass : Tetrapoda

Class : Amphibia

Order : Caudata or Urodela-Tail is present.

Type : *Salamandra*

Morphological Peculiarities of *Salamandra*

1. It is lizard like animal. They have teeth on both jaws. Vomerine teeth present.
2. Movable eyelids are present. Ear without ear drum or tympanic membrane.
3. Fore limbs with 4 digits and hind limbs with 5 toes.
4. Fore limbs and hind limbs are equal and weak.
5. It is black in colour, ovoviviparous and found in moist forest.
6. Body is brilliantly black with irregular yellow patches on limbs and back.
7. Head contains mouth prominent eyes and nostrils.
8. Species of Anderson's salamander found in marine water.
9. The skin of some species contains the powerful poison tetrodotoxin; these salamanders tend to be slow-moving and have bright warning coloration to advertise their toxicity.
10. They are capable of regenerating lost limbs and other damaged parts of body.
11. Body divisible into head, trunk and tail. The adult has no gills, eyelids are movable (Shinde *et al.*, 2009 and Mutakekar *et al.*, 2013).

1.14. *Chelone*

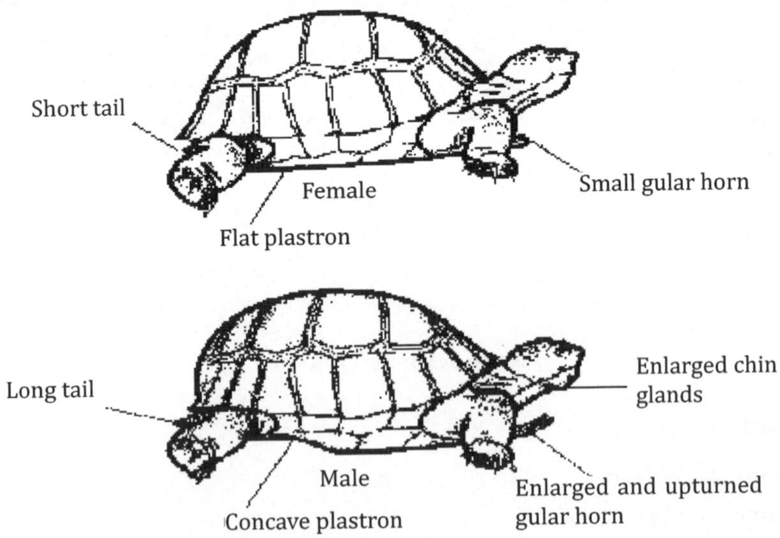

Figure 1.14: *Chelone*.

Classification

Phylum : Chordata

Subphylum : Vertebrata

Superclass : Tetrapoda

Class : Reptilia-Dry, scaly skin present. Tail present. Transverse cloaca. amniotic membrane present, cold blooded animals. They are cold blooded or poikilothermous or ectothermal vertebrates in which the body temperature is variable according to the environmental changes.

Subclass : Anapsida- Temporal cavities absent. Humerus and femur bones laterally extended.

Order : Chelonida or Testudines- Turtles and tortoises. Elliptical dorsoventrally flattened body. Convex carapace present.

Type : *Chelone*

Morphological Peculiarities of *Chelone*

1. It is commonly called sea turtle (*Chelone*) is aquatic (tortoise is semiaquatic) and dorsoventrally flattened animal. Retractile neck bends in a vertical 'S' curve.
2. Head partially withdrawn into carapace in turtle.
3. Head withdrawn into carapace in tortoise.
4. In male large gular horn and concave plastron present.
5. In male long tail than the female.
6. Ear drum present. Eyes with eye lids and nictitating membrane. Short tail present.
7. It is herbivorous and oviparous. Quadrate bone immovable.
8. The paddle like a pair for limbs (flippers) well developed for swimming.
9. The first digit fore limb is clawed. Hind limb toes are clawed.
10. The dorsal carapace is heart shaped (in tortoise ovoid carapace).
11. Forelimb is modified into paddle (in tortoise it is not modified).
12. Turtle is carnivorous (tortoise is herbivorous). The neck cannot be pulled into the shell. Jaw without teeth. Jaw with horny sheath and contains denticulated edges.
13. The dorsal surface of the turtle's head has a single pair of prefrontal scales. Its carapace (shell) is composed of five central scutes, four pairs of lateral scutes.
14. In above 30 °C favour development of females and below 30 °C produces males.

1.15. *Hemidactylus*

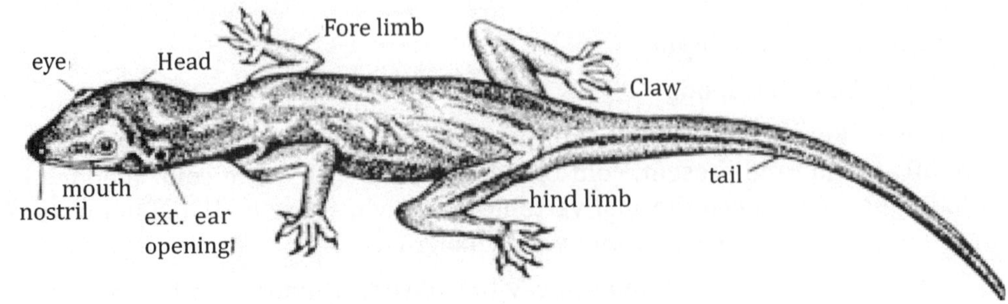

Figure 1.15: *Hemidactylus*.

Classification

Phylum : Chordata

Subphylum : Vertebrata

Supeclass : Tetrapoda

Class : Reptilia

Subclass : Lepidosauria: Diapsid reptiles. Two temporal cavities present. Lizards, snakes and sphenodons.

Order : Squamata- Scales present on the body. Lizards and snakes.

Suborder : Lacertilia- All lizards.

Type : *Hemidactylus*

Morphological Peculiarities of *Hemidactylus*

1. It is also known wall lizard or house gecko.
2. Body divisible into head, trunk and tail.
3. Head contains mouth, prominent eyes and nostrils.
4. The pupil is vertical. Eye lacks movable eyelids.
5. Fingers and toes are free, less webbed and dilated.
6. Dorsal scales granular or intermixed with large tubercles.
7. The digits and toes are provided with rounded specialized adhesive pads.
8. Tail with enlarged tubercles above and regenerates if broken.
9. Tail breaks off to escape from enemy. Nine to eleven lamellae under fourth toe.
10. The digits and toes underneath bear two rows of lamellae like compound leaf.
11. The wall lizard checks insect population. It is not poisonous.

12. Limb pads work on vacuum cup principle and help animal to move on ceiling.

1.16. *Chameleon*

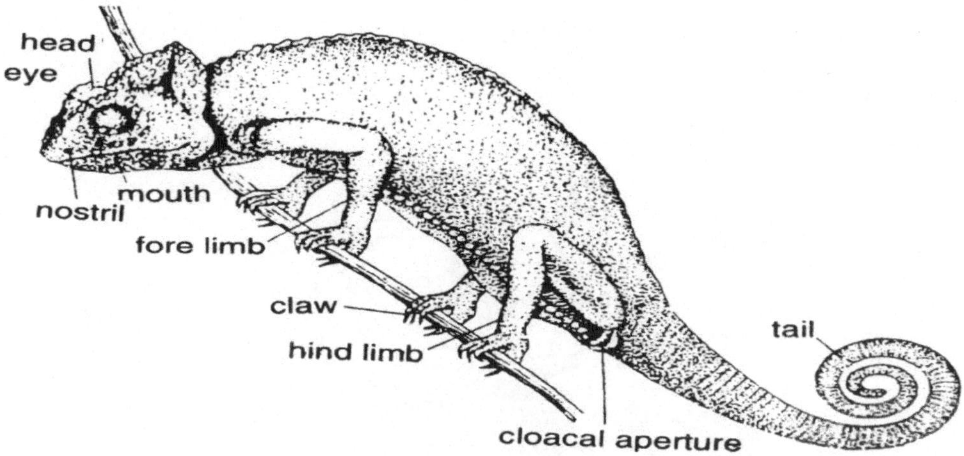

Figure 1.16: *Chameleon.*

Classification

Class : Reptilia

Subclass : Lepidosauria: Diapsid reptiles. Two temporal cavities present. Lizards, snakes and sphenodons.

Order : Squamata- Scales present on the body. Lizards and snakes.

Suborder : Lacertilia- All lizards.

Type : *Chameleon*

Morphological Peculiarities of *Chameleon*

1. Body divisible into head, trunk and tail.
2. Body is covered with scales.
3. Trunk is slightly bent. Body laterally compressed and covered with tubercles.
4. Eyes covered with a thick, granular lid, pierced by a transverse slit.
5. Large eyes present.
6. Small nostril present.
7. Cephalic hood present.
8. Long cylindrical, protrusible, sticky tongue is present.
9. Posterior part of head shows much elevated crest.

10. The toes and digits are opposed (2 versus 3) for grasping branches of trees.
11. Prehensile tail is present.
12. The color changes with suitable back ground.
13. It is the example of camouflage.

1.17. *Draco*

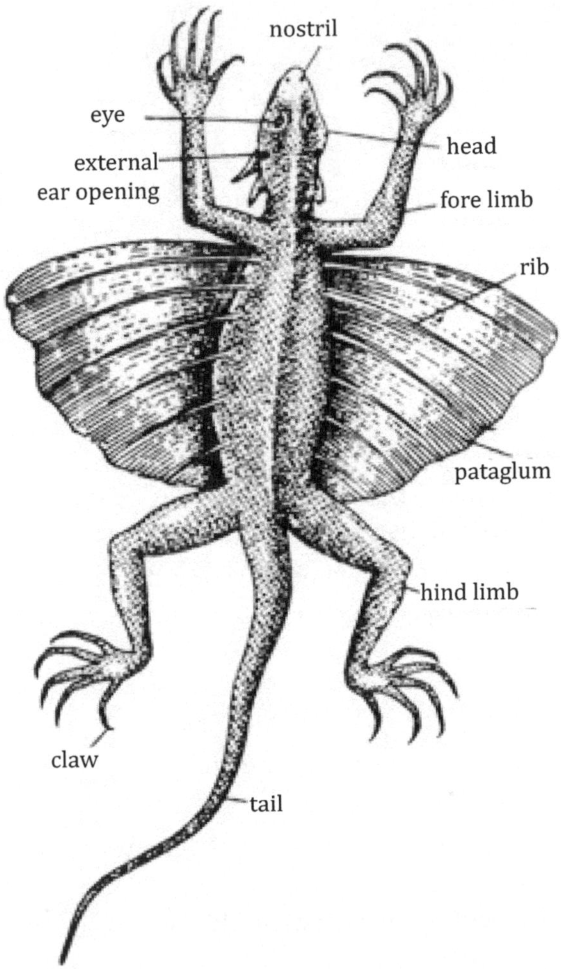

Figure 1.17: *Draco*.

Classification

Class : Reptilia

Subclass : Lepidosauria: Diapsid reptiles. Two temporal cavities present. Lizards, snakes and sphenodons.

Order : Squamata- Scales present on the body. Lizards and snakes.

Suborder : Lacertilia- All lizards.

Type : *Draco*

Morphological Peculiarities of *Draco*

1. It is commonly called flying lizard or flying dragon or agamid lizard.
2. Body divisible into head, trunk and tail.
3. It is dorsoventrally compressed for easy movement through foliage of tree.
4. Body is covered with scales.
5. It has large, bright colored, wing-like membranous folds of skin called patagium.
6. Patagium is supported by 5-7 extended ribs.
7. It is adapted for gliding (flying upto 200 feet). Tympanum is distinct.
8. Nostril opens at the top of snout. Upper lip scales are larger.
9. A prominent horn like tubercle present just behind the eyes.
10. Yellow colored gular pouch present below the throax and it is larger in male.
11. Three hooks present on the neck which help in taking support on tree branches.

1.18. *Crocodylus*

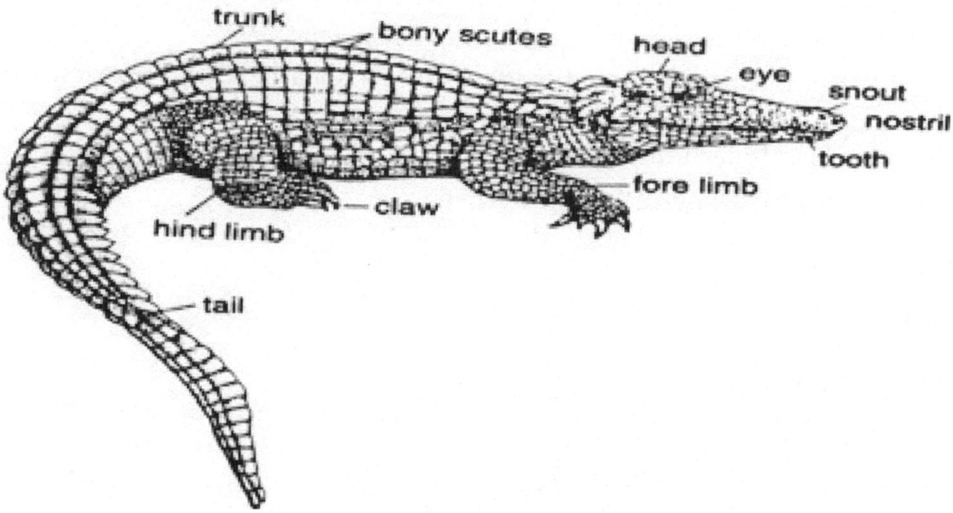

Figure 1.18: *Crocodylus*.

Classification

Phylum : Chordata

Subphylum : Vertebrata

Superclass : Tetrapoda

Class : Reptilia

Subclass : Archosauria-Presence of opening of fenestra into the skull.

Order : Crocodilia- Crocodiles and alligators. Thecodont teeth present.

Type : *Crocodylus*

Morphological Peculiarities of *Crocodylus*

1. It is commonly called mugger or marsh crocodile.
2. Body is covered with thick scales with dermal bony plates.
3. Eye, nostril and ear located on the top of the head.
4. It measures about 4 meters in length. It is dark olive brown with black spots.
5. Head bears long snout. Tongue flat and not protrusibel.
6. Fourth mandibular tooth fits into the notch in the upper jaw. Long snout present.
7. The tail is long and laterally compressed used in swimming. Heart is 4 chambered.
8. Upper and lower jaw contains 19 teeth on either side.
9. Ears are covered with movable flaps or skin. Teeth are polyphyodont type.
10. Fore limb bears five fingers and hind limb four toes.
11. Egg incubation period is 80 days. Life span is 35 to 75 years. It is carnivorous.
12. Cloacal aperture is longitudinal.
13. Single copulatory organ is present.

1.19. *Gavialis*

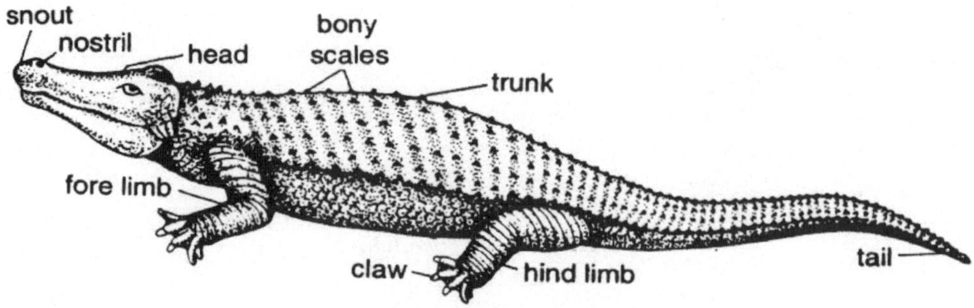

Figure 1.19: *Gavialis.*

Classification

Phylum : Chordata

Subphylum : Vertebrata

Superclass : Tetrapoda

Class : Reptilia

Subclass : Archosauria-Presence of opening of fenestra into the skull.

Order : Crocodilia- Crocodiles and alligators. Thecodont teeth present.

Type : *Gavialis*

Morphological Peculiarities of *Gavialis*

1. It is commonly called susar. It measures about 3 meters in length.
2. Body is covered with horny thick scales with dermal bony plates.
3. Head is triangular with broad, stout, short and rounded snout.
4. Body is covered by bony plates which are overlapped by horny epidermal plates.
5. Thecodont type of teeth. Fourth tooth of upper jaw is well developed.
6. Steel grey above and olive green at the sides. Heart is four chambered.
7. Tail laterally compressed used in swimming. Fore limb bears five fingers and hind limb four toes. Fingers and toes are webbed at the base.
8. The thoracic cavity is separated by diaphragm.
9. Urinary bladder is absent. Cloacal aperture is longitudinal.
10. Single copulatory organ is present.

1.20. Shrew

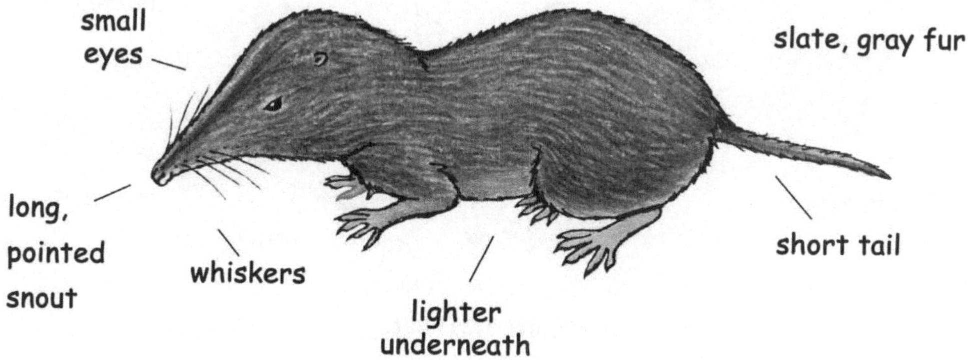

Figure 1.20: Shrew.

Classification

Phylum : Chordata

Subphylum : Vertebrata

Superclass : Tetrapoda

Class : Mammalia-Hairs and mammary glands present.

Subclass : Theria-Viviparous animal

Infraclass : Eutheria (Placentalia)- True placental animals. Teats present.

Order : Eulipotyphla- Hedgehogs and shrews. Snout present.

Type : Shrew

Morphological Peculiarities of shrew

1. It is called common shrew, *Sorex araneus.*
2. Body divisible into head, neck, trunk and tail.
3. It is gray at dorsal and pale at ventral side.
4. Small eyes with poor vision.
5. A pair of incisors long and sharp.
6. It is insectivorous and feeds on seeds, insects, nuts and worms.
7. Soft and dense fur present.
8. It is found in dense vegetation.
9. It has good senses of hearing and smell.
10. Some species of shrews are venomous.

1.21. Bat

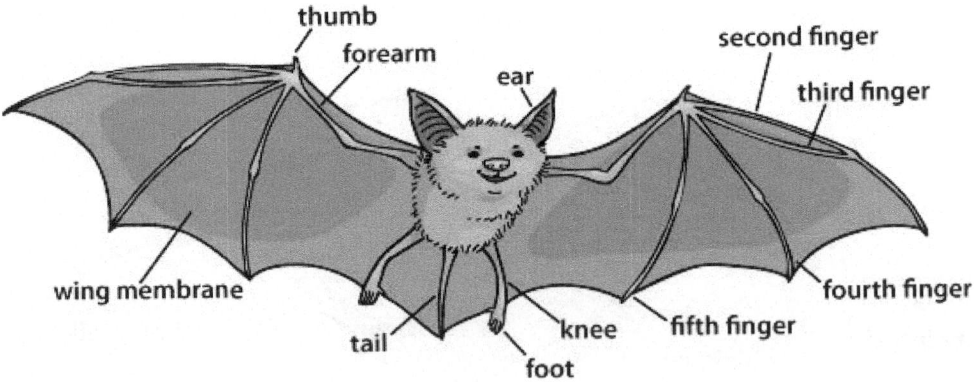

Figure 1.21: Bat.

Classification

Class : Mammalia

Subclass : Theria

Infraclass : Eutheria

Order : Chiroptera-Fore limbs modified for flight.

Type : Bat

Morphological Peculiarities of bat

1. It is called bat or flying fox, *Pteropus giganteus.*
2. Body divisible into head, neck, trunk and tail.
3. Soft brown fur present on the skin. It has wing span up to 150 cm.
4. Body is dark brown and shoulder golden yellow color.
5. Wing is formed by a fold of skin called patagium.
6. Patagium is supported by 2nd and 5th fingers. 1st and 2nd fingers bears claws.
7. Except thumb all the fingers are elongated.
8. Hind limbs and tail also included into patagium.
9. Hind limb bears 5 clawed digits. Snout is short. Pinna well developed.
10. Testes abdominal. Eyes small. Snout is elongated.
11. Hairs are present on the body. It is nocturnal animal.
12. It is insectivorous, frugivorous and sanguivorous animal.
13. They produce ultrasonic sound waves which help in flight.

1.22. Squirrel

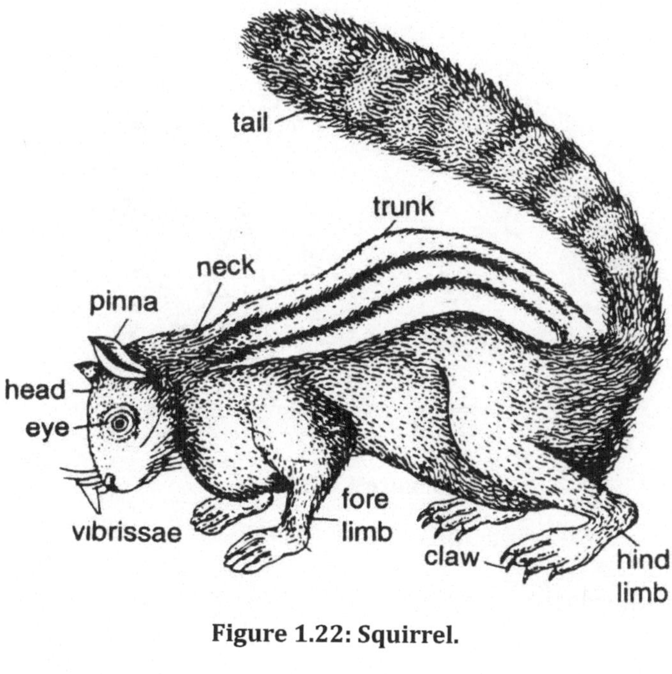

Figure 1.22: Squirrel.

Classification

Phylum : Chordata

Subphylum : Vertebrata

Superclass : Tetrapoda

Class : Mammalia

Subclass : Theria

Infraclass : Eutheria

Order : Rodentia-Gnawing animals.

Type : Squirrel

Morphological Peculiarities of Squirrel

1. It is called Indian three striped palm squirrel, *Fundabulus palmarum*.
2. Body divisible into head, neck, trunk and tail.
3. Three white stripes alternate with black present on the dorsal surface.
4. Ventral side and limbs are covered by small grey hairs.
5. Large external ears (pinnae) used as tactile organ.
6. Head bears eyes, nostrils and mouth.
7. Elongated snout shows small vibrissae.
8. Tail is elongated and bushy. Clawed toes and digits.

1.23. Loris

Figure 1.23: Loris.

Classification

Phylum : Chordata

Subphylum : Vertebrata

Superclass : Tetrapoda

Class : Mammalia

Subclass : Theria

Infraclass : Eutheria

Order : Primate

Type : Loris

Morphological Peculiarities of Loris

1. Loris, *Nycticebus kayana* is an arboreal primate.
2. Body divisible into head, neck, trunk and tail.
3. The body is covered by brownish wooly fur.
4. Loris is tailless animal.
5. Snout present. Conical external ear present.
6. They eat fruits and small animals.
7. Small apertures of nostrils.
8. Some toes are clawed and some toes are with nails.
9. Head is fox like shows pointed muzzle.
10. Eyes bulging and closely located at the front side of face.
11. External ear or pinna is short and conical (Ghose and Manna, 2016).

2. STUDY OF SNAKES

Aim: Characters identifying venomous snakes and non-venomous snakes

Characters Identifying Venomous Snakes

1. Head is triangular.
2. Small scales are present on the dorsal side of head.
3. Eyes with elliptical pupil.
4. Pit (useful in thermal sensation) is present just below the nostril.
5. Third supra labial shield touches to the eye and nostril (cobra, coral snake).
6. A pair of large expanded elevations is present at the posterior side of head which represent the position of venom glands.
7. Hood and spectacle mark (cobra), black and white cross bands (krait) present.
8. Arrow mark and heat sensing pit on the head (viper) present.

9. The middle row of scales on the back called vertebrals may be larger that other.
10. A pair of poison fangs is present at the upper jaw.
11. Dorsal and ventral scales are differentiated from each other.
12. In poisonous snake bite many punctures occur along with the fangs punctures.
13. Single row of scales at the ventral side of tail.
14. It swims at the surface of water.
15. Flow of blood occurs from wound. Swelling of wound occurs.
16. Ventral side of the lower jaw has fourth infra-labial shield larger than others.
17. Some examples are ghonus, phoorsa, manyar, nag, sea snakes, coral snake, king cobra, tiger snake, green pit viper, Inland taipan and black mamba.

Characters Identifying Non-Venomous Snakes

1. Head is rounded or 'U' shaped.
2. Large scales are present on dorsal side of the head.
3. Eyes with rounded pupil.
4. Pit absent just below the nostril.
5. Third supra labial shield does not touch to the eye and nostril.
6. A pair of large expanded elevations is absent at the posterior side of head.
7. Hood, spectacle mark, black and white cross bands absent.
8. Arrow mark and heat sensing pit absent on the head.
9. The middle row of scales on the back called vertebrals are similar with other.
10. A pair of poison fangs is absent at the upper jaw.
11. Dorsal and ventral scales are not differentiated from each other.
12. In non-poisonous snake bite many small punctures occur.
13. Two rows of scales at the ventral side of tail.
14. It swims in the deep water.
15. Flow of blood stops due to natural clotting. Swelling of wound does not occur.
16. Ventral side of the lower jaw has fourth infra-labial shield similar as in others.
17. Rat snake, python, water snake, sand boa (durkya ghonus), earth boa (mandval), blind snake, *green keelback, worm snake, banded racer,* gunther's racer, striped keelback, banded kutri

Venomous Snakes

1.24. Russell's Viper

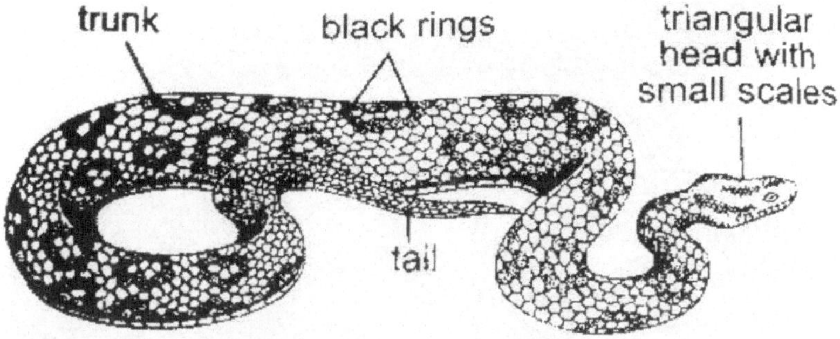

Figure 1.24: Russell's Viper.

Classification

Class : Reptilia

Subclass : Lepidosauria: Diapsid reptiles. Two temporal cavities present. Lizards, snakes and sphenodons.

Order : Squamata- Scales present on the body. Lizards and snakes.

Suborder : Ophidia. Snakes

Type : Russell's viper

Characters Identifying Venomous Snake in Russell's Viper

1. It is called ghonus, *Vipera russeli.*
2. It is a venomous snake. It is pitless viper. Ventral surface yellowish.
3. Head is distinct from neck and covered with small scales and few large shields.
4. A pair of elevations is present at the posterior side of head representing the position of venom glands. Large nostrils. Golden eyes with elliptical pupil.
5. Pale brown color above and yellow below having scattered dark spots.
6. Three longitudinal series of black light edged rings, which sometimes encircle reddish spots. A pair of big solenoglyphous fangs is present at the upper jaw.
7. Head is triangular and covered with very small imbricate scales on the upper surface. Snout is angular and the nasal opening is quite prominent.
8. It has haemotoxic venom. Scales are sharply keeled.
9. When it bites there is tremendous pain and swelling, irritation and death due to heart failure.

1.25. Saw Scaled Viper

Figure 1.25: Saw Scaled Viper.

Characters Identifying Venomous Snakes in Saw Scaled Viper

1. It is called phoorsa, *Echis carinatus.* It is pit viper.
2. It is a venomous snake. Small scales are present on the head.
3. A pair of elevations is present at the posterior side of head which represent the position of venom glands.
4. Head is short wide, pear shaped and distinct from the neck.
5. A pair of solenoglyphous fangs is present at the upper jaw.
6. Snout is short and rounded. Eyes with elliptical (vertical) pupil.
7. The crown is covered with small, irregular imbricate scales.
8. The dorsal body scales are keeled.
9. White arrow mark on the head behind eyes.
10. When it bites there is tremendous pain and swelling, irritation and heart failure.
11. It has haemotoxic venom.

1.26. Common Krait

Characters Identifying Venomous Snakes in Common Krait

1. It is called krait, *Bungarus fasciatus.* It is a venomous snake.
2. Head not distinct from neck and small scales are present on the head.

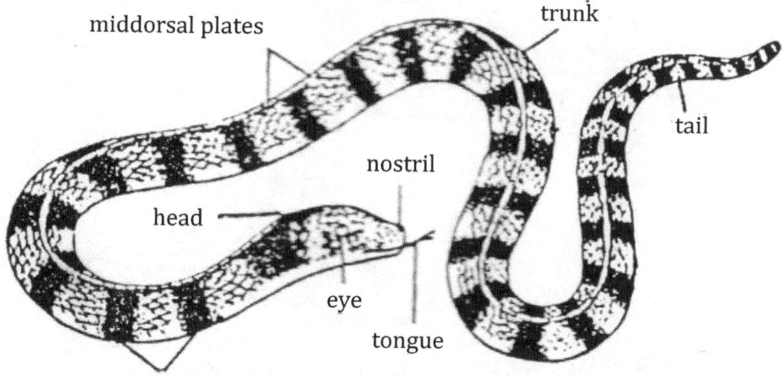

Figure 1.26: Common Krait.

3. It measures about 1.5 to 2.5 meters in length. Mouth small.
4. Dark cross bands are present on the body. Bands absent on head.
5. Single row of large hexagonal scales present on the middorsal side of body.
6. A pair of elevations is present at the posterior side of head which represent the position of venom glands.
7. A pair of small proteroglyphous fangs is present at the upper jaw.
8. Scales beyond the anal region are in a single row.
9. When it bites there is tremendous pain and swelling and continuous blood flow from wound, even there is internal bleeding in organs.
10. It has neurotoxic venom.
11. Eyes with elliptical (vertical) pupil.
12. Scales in 15 rows.

1.27. Indian Cobra

Characters Identifying Venomous Snakes in Indian Cobra

1. It is called nag, *Naja naja*. It is venomous snake.
2. It measures about 4 to 5 meters in length.
3. Small scales are present on the triangular head.
4. It has neurotoxic venom. It is wheatish in color.
5. In hibernation it becomes golden but on exposure to sunlight it changes to brown.
6. Body is brown or blackish.
7. Body is covered with smooth oblique scales arranged in 15-25 rows.

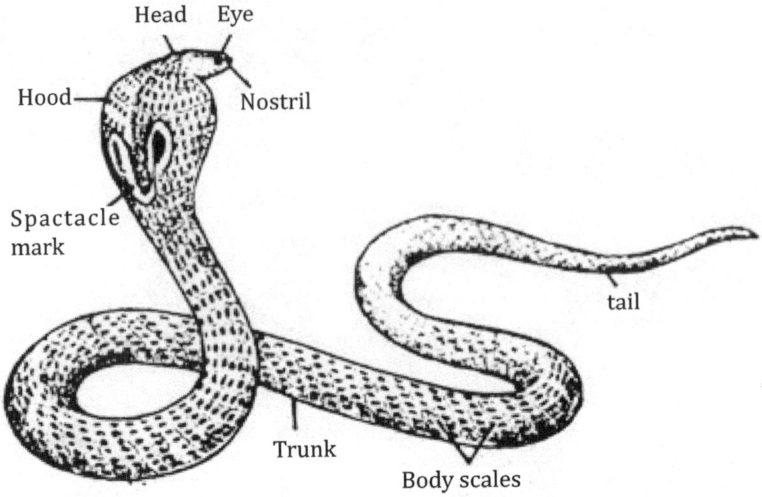

Figure 1.27: Indian Cobra.

8. A pair of elevations is present at the posterior side of head which represent the position of venom glands. Poison fangs are present.
9. The expansion of the neck and the cervical ribs forms hood with spectacle mark.
10. The upper surface of the hood bears spectacle mark (figure of ten).
11. Spectacle mark is also called binocellate mark (Maharashtra), monocellate mark (Bengal) and non-cellate (Gujarat, M. P.).
12. Eyes with elliptical (vertical) pupil.
13. Third supra-labial shield touches to eye and nostril.
14. Tail pointed and cylindrical.
15. A pair of proteroglyphous fangs is present at the upper jaw. It feeds on frogs, lizards, rats and other snakes.

1.28. Sea Snake

Characters Identifying Venomous Snakes in Sea Snake

1. It is called *Hydrophis ornatus*. It is a venomous snake.
2. Small scales are present on the triangular head.
3. Body round and tail is laterally compressed.
4. Head and neck slender. Head shields unbroken. Nostrils are dorsal.
5. Eyes small with elliptical (vertical) pupil. Ventral scales are very small.
6. A pair of elevations is present at the posterior side of head which represent the position of venom glands. It has neurotoxic venom.

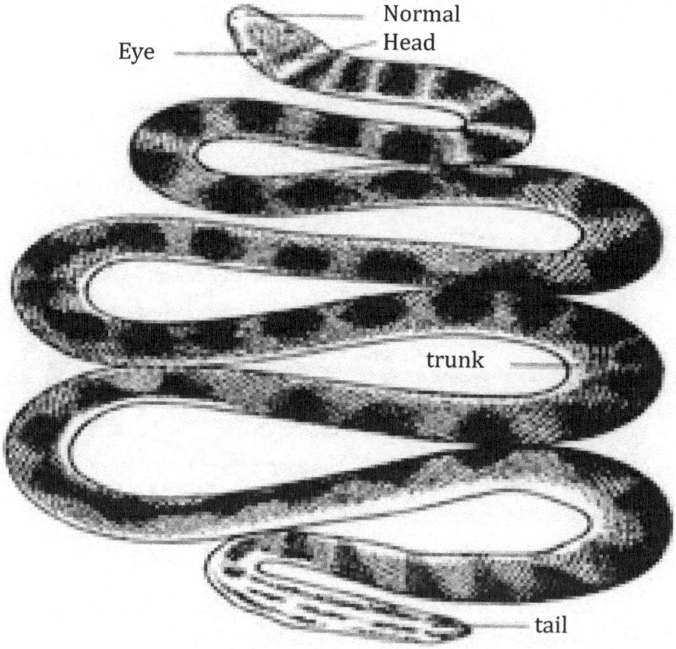

Figure 1.28: Sea Snake.

7. It measures about 1 meter in length. It is carnivorous. It is viviparous.
8. Colour is dark olive green above with yellowish cross bars and white below.
9. Poison fangs are present.
10. Dark bands present on the body.
11. Ventral scales are small and differentiated from dorsal scales.
12. A pair of proteroglyphous fangs is present at the upper jaw.
13. It feeds on frogs, lizards, rats and other snakes.

Non-Venomous Snake

1.29. Rat Snake

Characters Identifying Non-venomous Snakes in Rat Snake

1. It is called Indian dhaman, *Ptyas* (*Zamenis*) *mucosa*. It is a non-venomous snake.
2. It measures about 3 to 4 meters in length. Large scales are present on the head.
3. It grows to a length of more than two meters. Head is rounded or 'U' shaped.

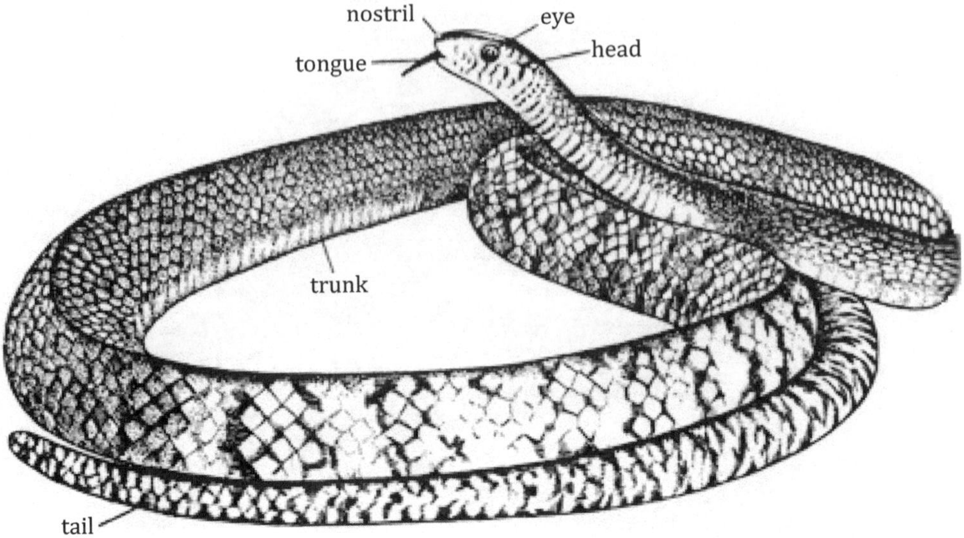

Figure 1.29: Rat Snake.

4. Colour is black brown above with darker cross bands; at least on the posterior part; below yellowish-white. Head is distinct from the neck.
5. It is fast moving, diurnal and semi-arboreal.
6. Eyes large with round pupil.
7. Fangs absent. Maxillary teeth are about 22-28.
8. Dorsal ridge is present on the mid dorsal line.
9. After biting coils around the victim firm by long prehensile tail.
10. Tail is long and carries two rows of scales ventrally.
11. Scales rows 16-17.
12. It feeds on rodents like rat. It is oviparous.
13. Fourth and fifth supra labials are touching the eye.

1.30. Checkered Keel Back

Characters Identifying Non-venomous Snakes in Checkered Keel Back

1. It is called Asiatic water snake, *Xenochrophis* (*Natrix*) *piscator.*
2. It is a non-venomous snake. Nostril situated on top of head.
3. It measures about 1 to 2 meters in length.
4. Presence of two black streaks running behind the eye on each side.
5. Large scales are present on the 'U' shaped head. Body is stout.
6. Eyes with rounded pupil.
7. Body is stout. Small tail.

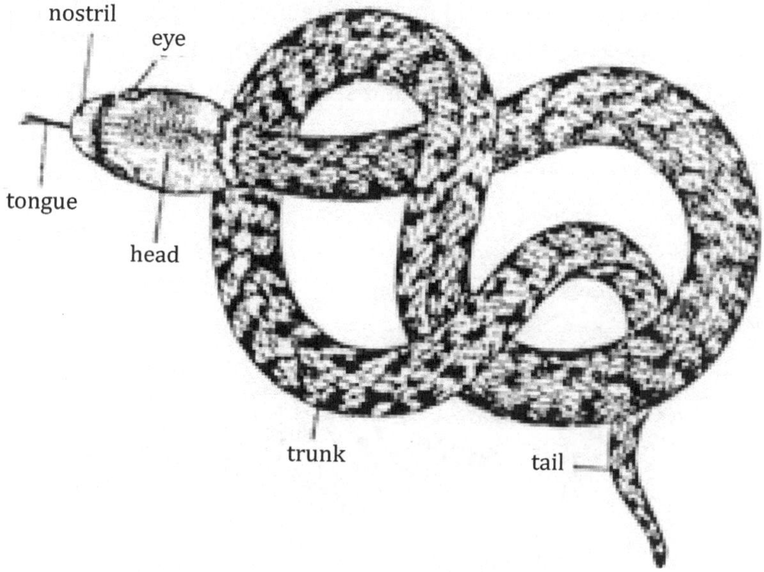

Figure 1.30: Checkered Keel Back.

8. Only one preocular scale below the both eyes.
9. Body stout and head is flat.
10. Scales rows 15-19.
11. Maxillary teeth 22-28.
12. Fangs absent.
13. Large scales are present on the head.
14. Colour is olive with black spots arranged in the chess board manner.
15. Scales are in 19 distinct rows.
16. Ventral scales are rounded and subcaudal scals are divided.
17. The black spots on the back are set in a chess board pattern and hence its common name "Checkered keelback" (Ghose and Manna, 2016).

Study of Birds

Adaption for Flight in Birds

1. Beaks modified for feeding like seed, fruit eating and fish catching beaks.
2. Legs modified for locomotion in different habitats like climbing, swimming feets.
3. Pneumatic bones for reduction in weight and flight.
4. Monocondylic skull for easy and quick movement of head to see objects.
5. Long, mobile neck for all round sight. Crop used in temporary storage of food.

6. Well developed feathered for sustenance in air and flight.
7. Tail feathers used for regulation of flight. Bipedal and feathered animals.
8. Limbs adapted for perching and spontaneous grip on support.

1.31. Sparrow

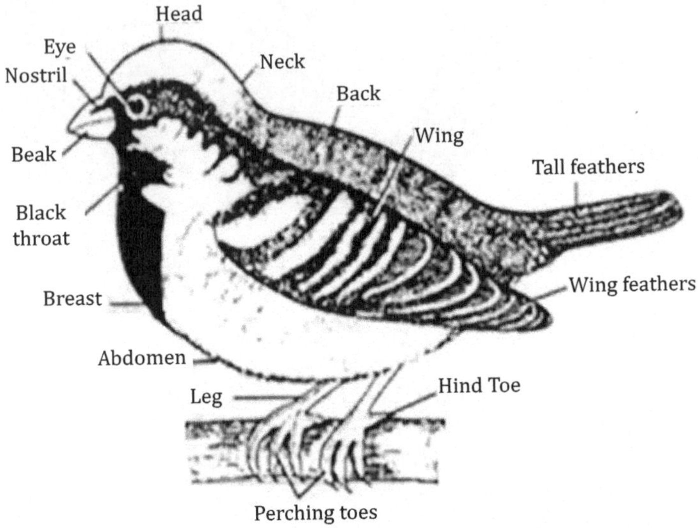

Figure 131: Sparrow.

Phylum : Chordata

Subphylum: Vertebrata- Vertebral column is present.

Superclass : Tetrapoda- Forelimbs and hindlimbs are present.

Class : Aves-Bipedal, feathered, first warm blooded animals.

Subclass : Neornithes- Modern birds. Caudal vertebrae fused to form a pygostyle.

Superorder : Neognathae- Sternum with keel. Wings well developed.

Order : Passeriformes-Three toes in front and one behind, adapted for perching.

Type: Sparrow

General Characters of Sparrow

1. It is called house sparrow, *Passer domesticus*.
2. It has short, stout, slightly curved seed eating beak.
3. It grows up to a length of 12.5 cm. Female is ash-white in color.
4. Male is earthy brown with blackish throat and breast and white abdomen.
5. Back streaked with black. Pale brown legs.
6. Perching feet present. Crown of the head ashy grey.

7. It feeds on seeds and gains fallen on the ground.
8. It has 2nd, 3rd and 4 th toes at the front and 1st toe hallus at the back.

1.32. Kingfisher

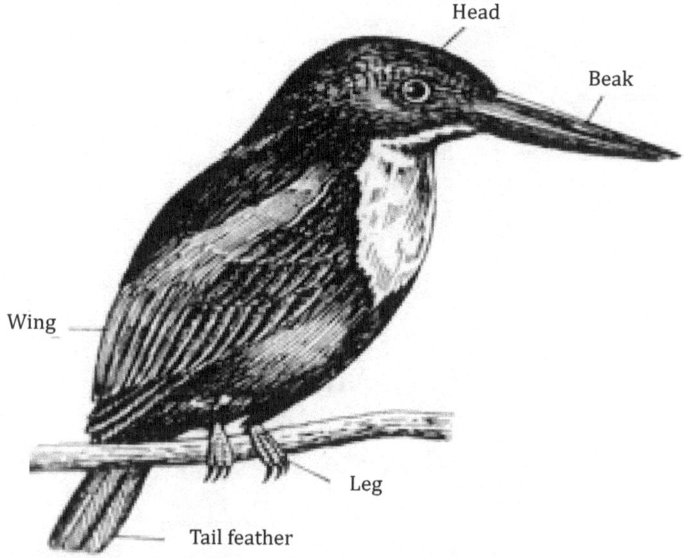

Figure 1.32: Kingfisher.

Class : Aves- Feathered animals. Forelimbs are modified into wings.

Subclass : Neornithes- Modern birds. Caudal vertebrae fused to form a pygostyle.

Superorder : Neognathae- Sternum with keel. Wings well developed.

Order : Coraciformes- Crow like bird. 2st and 3nd toes fused at the base.

Type : Kingfisher

General Characters of Kingfisher

1. It is called common king fisher, *Alcedo atthis bengalensis.*
2. It is purple blue and green in color. It is about 18 cm long.
3. Black beak and reddish legs. Fish catching beak. White color on the neck.
4. Climbing feet with 1st and 4 th toes at the back; 2nd, 3rd toes on the front side.
5. In king fisher 2nd and 3rd toes fused throughout the length called syndactyle feet.
6. Two toes are forwardly directed and two toes are backwardly directed.
7. Short stumpy tail. Crest on the head present.

1.33. Woodpecker

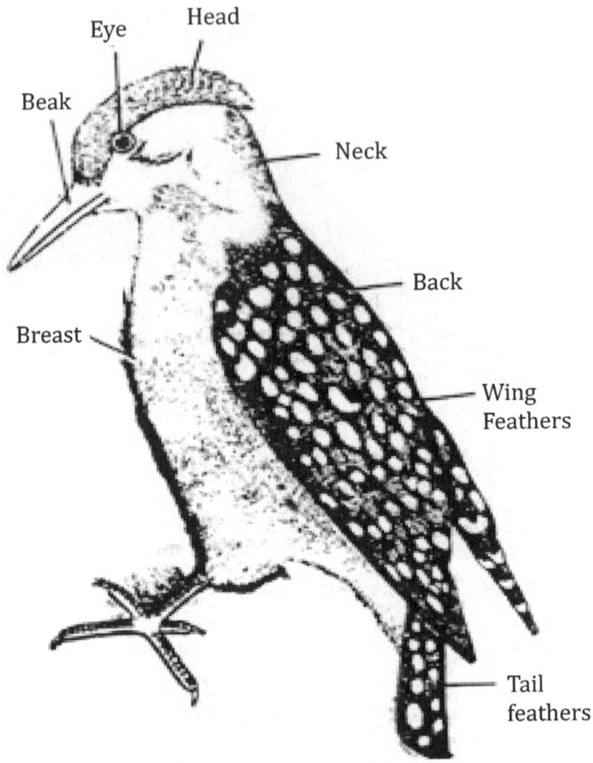

Figure 1.33: Woodpecker.

Class : Aves- Bipedal and feathered animals. Warm blooded animals.

Subclass : Neornithes- Modern birds. Caudal vertebrae fused to form a pygostyle.

Superorder: Neognathae- Sternum with keel. Wings well developed.

Order : Piciformes- beak strong and stout.

Type : Woodpecker

General Characters of Woodpecker

1. It is commonly called golden-backed woodpecker. Crown lies on the head.
2. Upper side is golden yellow and black.
3. The lower side is white and streaked with black.
4. The beak is long, stout and pointed and wood pecking.
5. Tail feathers with pointed tips.
6. Found in villages and highly wooded area.
7. Wood-chiselling beak present.

8. Climbing feet present. Tongue protrusible.
9. Feet zygodactyls with 1st and 4th toes at the back and 2nd, 3rd on the front side.

1.34. Kite

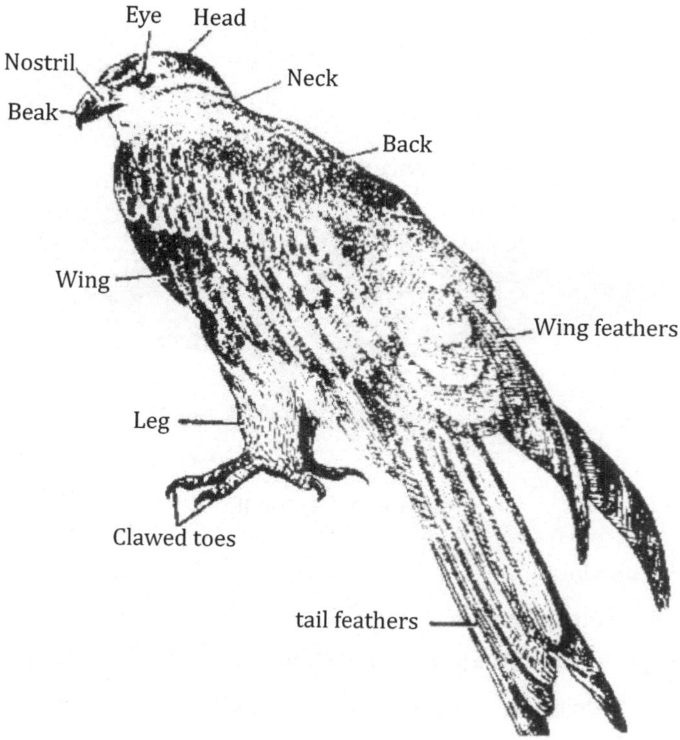

Figure 1.34: Kite.

Class : Aves- Feathered animals. Forelimbs are modified into wings.

Subclass : Neornithes- Modern birds. Caudal vertebrae fused to form a pygostyle.

Superorder : Neognathae- Sternum with keel. Wings well developed.

Order : Falconiformes, beak strong and hooked.

Type : Kite

General Characters of Kite

1. It is brown in colour. Strong claws present.
2. Lives near the human habitation. Raptorial feet present.
3. It is carnivorous and food chiefly includes lizards, mice, young birds and snakes.

4. Piercing and tearing beak present. Toes provided with ventral bulbs or tylari.
5. It has 2nd, 3rd and 4th toes at the front and 1st toe hallus at the back.

1.35. Duck

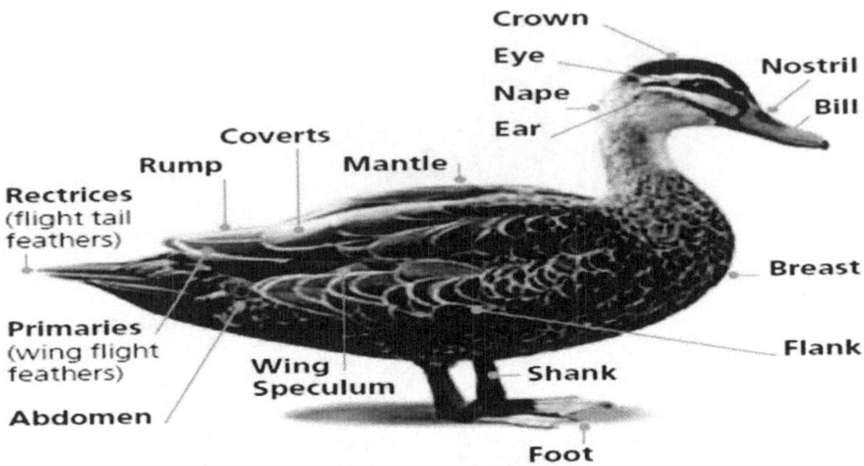

Figure 1.35: Duck.

Phylum : Chordata

Superclass : Tetrapoda- Forelimbs and hindlimbs are present.

Class : Aves- Feathered animals. Forelimbs are modified into wings.

Subclass : Neornithes- Modern birds. Caudal vertebrae fused to form a pygostyle.

Superorder : Neognathae- Sternum with keel. Wings well developed.

Order : Anseriformes, flat beak and webbed toes.

Type : Duck

General Characters of Duck

1. It is aquatic bird. Body divisible into head, neck, trunk and tail.
2. Beak is broad and provided with plates called as lamellae.
3. Margins of the beak contain many transverse horny ridges or lamellae.
4. Tongue is fleshy. Male brighter than female.
5. Legs are short and feet are webbed.
6. Tail feathers are short. Fleshy tongue present.
7. Water and mud straining beak present.
8. Swimming and paddling feet present. Three toes are included in the web.
9. It has 2nd, 3rd and 4th toes at the front and 1st toe hallus at the back.

1.36. Fowl

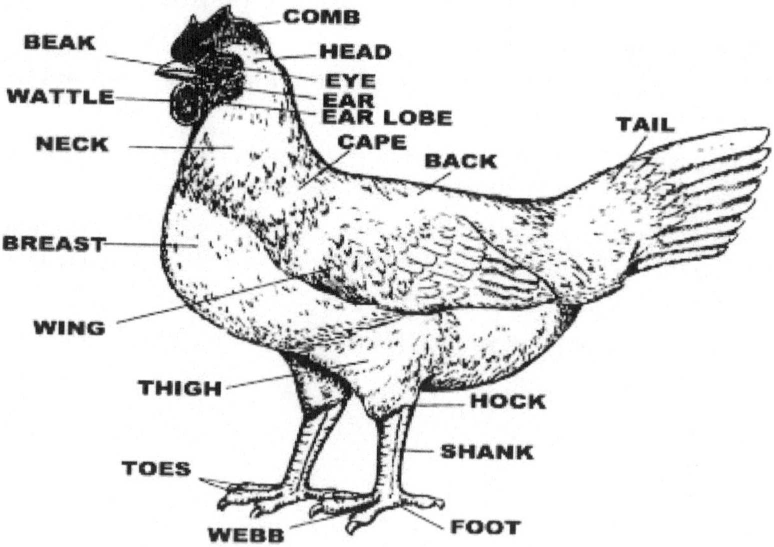

Figure 1.36: Fowl.

Phylum : Chordata

Superclass : Tetrapoda- Forelimbs and hindlimbs are present.

Class : Aves- Feathered animals. Head small, Teeth absent.

Forelimbs are modified into wings. Lungs with air sacs. Tail vertebrae few and compressed. Exoskeleton in the form of feathers and beak.

Subclass : Neornithes- Modern birds. Caudal vertebrae fused to form a pygostyle. Sternum well developed with keel. Caudal vertebrae 13 or less.

Superorder: Neognathae- Sternum with keel. Tail short with terminal pygostyle. Second finger longer. Feathers with barbs, barbules and hooklets.

Order : Galiciformes, 2nd, 3rd and 4th toes are forwardly directed and 1st toe is backwardly directed.

Type : Fowl

General Characters of Fowl

1. Terrestrial game birds with short and powerful flight.
2. Beak is short and it is seed eating type.
3. Long and mobile neck. Single occipital condyle.
4. Hind limbs bear claws and scales.
5. Feet usually massive adapted for scratching and running.
6. Graminivorous. Scratching and walking feet present.

7. It has 2nd, 3rd and 4 th toes at the front and 1st toe hallus at the back.
8. A chicken is differentiated from most other birds.
9. It has a comb and two wattles.
10. The comb is the red appendage on the top of the head and the wattles are the two appendages under the chin.
11. Good layers have clean-cut, strong, refined heads while poor layers have coarse, meaty/thin, blocky, weak looking head.
12. A good layer has an abdomen that is deep and soft, easily pliable without body fat accumulation by probing fingers while a poor layer has a hard and shallow abdomen.
13. Bony spur is present in cock (Mutakekar *et al.*, 2011).

Chapter 2

Brain of Fowl

Dissection of Brain of Fowl

Location

Brain is a hollow organ of central nervous system. It is located into the bony cranium.

Procedure

1. Remove the skin and muscle from the dorsal side of head of fowl.
2. With the help of scalpel breaks cranium from Lt. side of head and expose brain.
3. Remove the protective membranes (menings) outer duramater and inner arachnoid membrane. Clean the brain and kept in the watch glass with water.

Structure and Functions of Brain

The brain is mainly divided into three parts such as-

A. Forebrain or Prosencephalon
B. Midbrain or Mesencephalon
C. Hindbrain or Rhombencephalon

A. Forebrain

The parts of the forebrain are given below:

1. Olfactory lobes

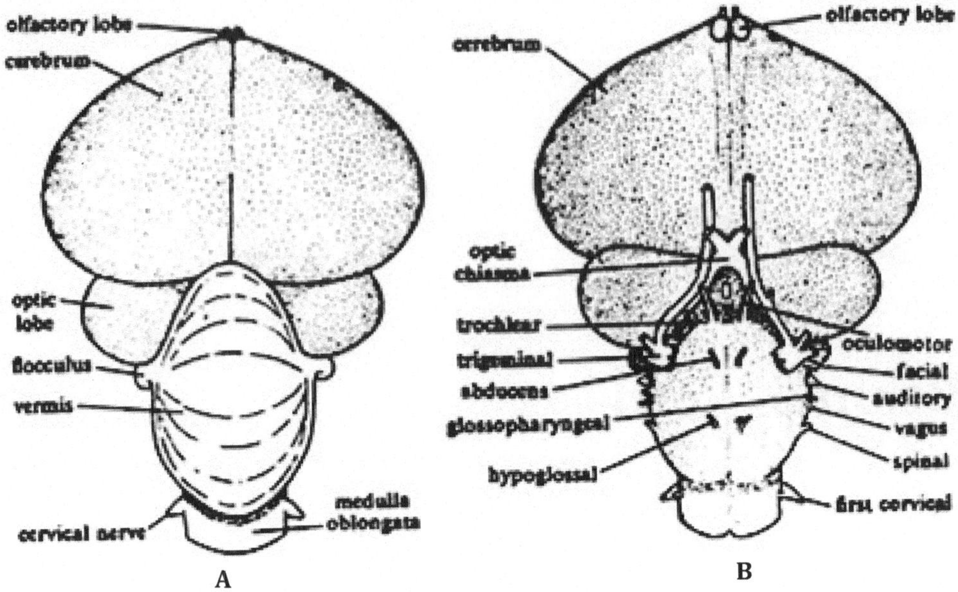

Figure 2.1: Brain of Fowl.
A: Dorsal view; B: Ventral view.

2. Cerebral hemispheres
3. Diencephalon
4. Anterior choroid plexus

1. *Olfactory Lobes*

A pair of very small olfactory lobes is directly attached to the cerebral hemispheres. Olfactory lobes enclose cavities called rhinocoels.

Function

1. Sense of smell. It is not well developed.

2. *Cerebral Hemispheres*

A pair of large rounded cerebral hemispheres is present just above the optic lobes. Two cerebral hemispheres are separated by median dorsal fissure. Cerebral hemispheres are smooth and folds or fissures are absent which are present in mammals. The ventrolateral walls of cavities are very thick and well developed called corpora striata. They enclose cavities called paracoels or lateral ventricles or Ist and IInd ventricles. The cavity of brain is called ventricle.

Function

1. It is the center of all voluntary activities like walking, running and flying.
2. It is the center of memory, intelligence and will power.

3. *Diencephalon*

It is unpaired part of forebrain. It is covered by cerebral hemispheres. It is not visible from dorsal side. It encloses a cavity called diocoel or IIIrd ventricle. Pineal body or epiphysis is present at the dorsal side of diencephalon and pituitary gland or hypophysis is present at the ventral side of diencephalon. Pineal body secretes melatonin hormone for normal functioning of neurons and induction of sleep. Pituitary gland is the master endocrine gland of the body and secretes ACTH, TSH, GnH, GH, PL and MSH hormones.

Function

1. It is the main center of autonomic nervous system.
2. It helps in the perception of metabolism, reproduction, pains and temperature.

4. *Anterior Choroid Plexus*

It is non-nervous, vascular and muscular part of forebrain.

Function

1. It secretes cerebrospinal fluid (C.S.F.).
2. Cerebrospinal fluid (C.S.F.) absorbs mechanical shock and helps in the transport of O_2, CO_2, food and excretory substances.

B. Midbrain

It includes following parts such as:

1. *Optic Lobes*

A pair of large, oval optic lobes is located at the lateral sides of cerebral hemispheres. Two well developed optic lobes are present in case of the bird hence called corpora bigemina. They enclose cavities called optocoels. Two optocoels are connected to the diocoel by a ventricle passage called iter or aqueductu of sylcius. Small passage of ventricle in between paracoels and diocoel called foramen of manro.

Function

1. Sense of sight or vision.

2. *Crura Cerebri*

It is the longitudinal band of nerve fibres. It is seen beneath the optic lobes.

It connects diencephalon with medulla oblongata means forebrain to hindbrain.

Function

1. Transport of nerve impulse from forebrain to hindbrain and vice versa.

C. Hindbrain

The parts of the hindbrain are given below:

1. Cerebellum
2. Medulla oblongata
3. Posterior choroid plexus

1. Cerebellum

It is well developed and relatively larger than the mammals. It is divided into three lobes such as large median lobe called vermis and two small lateral lobes or floccular lobes. Iter ventricle lies in between fore brain and hin brain.

Function

1. Sense of equilibrium (balance).

2. Medulla Oblongata

It is covered by cerebellum. It tapers towards spinal cord. It encloses a cavity called metacoel or myelocoel or IVth ventricle.

Function

1. It is the center of all involuntary activities like respiration, heart beats, peristalsis and opening and closing of eye lids.

3. Posterior Choroid Plexus

It is non-nervous, vascular and muscular part of hindbrain.

Function: Secretion of cerebrospinal fluid (C.S.F.) that absorbs mechanical shock and helps in the transport of O_2, CO_2, food and excretory substances.

Hyoid Apparatus of Fowl

Aim: Temporary Preparation of Hyoid Apparatus of Fowl.

Materials

Head of fowl.

Procedure

1. Take a head of fowl and locate the hyoid apparatus from lower jaw.
2. Take cut from the base of beak by protecting the posterior cornua.
3. Remove tongue with hyoid apparatus.
4. Trace posterior cornu, urohyal, basihyal and entoglossus by removing muscles and sheaths by forcep and scissor.
5. Treat with 10 per cent KOH, wash with water and mount with glycerine or water.

Location

Hyoid apparatus is present in the lower jaw just below the tongue of fowl.

Structure of Hyoid Apparatus

1. Hyoid apparatus is a frame work of bones and cartilages.
2. It is present in amphibians, reptiles and birds.
3. It has three arrow shaped parts called entoglossus, basihyal and urohyal.
4. Urohyal bears cartilage at the end.
5. A pair of anterior and posterior horns (cornua) is present.
6. Each posterior horn has anterior ceratobranchial and posterior epibranchial.
7. Epibranchial gradually narrow, pointed and cartilaginous at the tip.

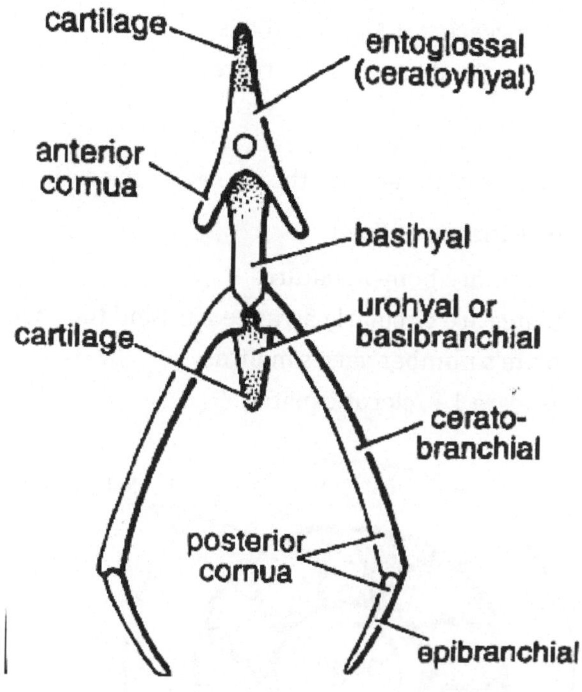

Figure 2.2: Hyoid Apparatus of Fowl.

Function

a. It gives support to the tongue.
b. It has taxonomic importance (Ghose and Manna, 2016).

Sclerotic Plates of Fowl

Aim: Temporary preparation of sclerotic plates of fowl.

Materials

Head of fowl, Alizarin stain and glycerine.

Procedure

1. Take a head of fowl and remove the eyeball from orbital cavity.
2. Cut the eyeball into two halves.
3. Take round of upper halve around of corneal layer.
4. Remove muscles by blade from both sides of cornea.
5. Treat with 10 per cent KOH and wash with water.
6. Kept in watch glass for staining with Alizarin red.
7. Mount with glycerine in the two slides.
8. Observe under the dissecting microscope.

Location

Sclerotic plates are present around the cornea of sclerotic layer.

Structure of Sclerotic Plates of Fowl

1. Sclerotic plates are bony in nature.
2. Sclerotic plates are circularly arranged around the cornea.
3. Sclerotic plate's number varies in birds.
4. In fowl there are 14 sclerotic plates.

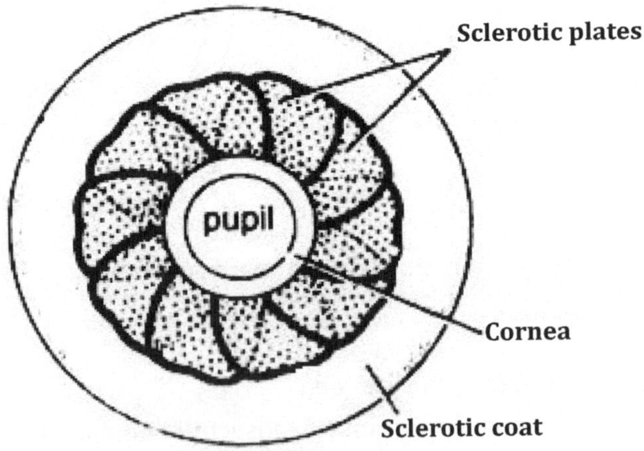

Figure 2.3: Sclerotic Plates of Fowl.

Function

1. They give protection to the eyeball.
2. They maintain the shape of eye ball.
3. They have taxonomic importance (Ghose and Manna, 2016).

Pecten of fowl

Aim: Temporary preparation of pecten of fowl.

Materials

Head of fowl, Eosin stain and glycerine.

Procedure

1. Take a head of fowl and remove the eyeball from orbital cavity.
2. Make incision in the equatorial plane of eyeball.
3. Cut the eyeball into two halves by scissor.
4. In posterior halve locate the pecten.
3. Separate the pecten with the help of blade.
4. Kept in watch glass for staining with eosin stain and mount with glycerine.
5. Observe under the dissecting microscope.

Location

Pecten is present at the blind spot region where in optic nerve enters into the eyeball. It is present at the tip of entry of the optic nerve into the eye-ball.

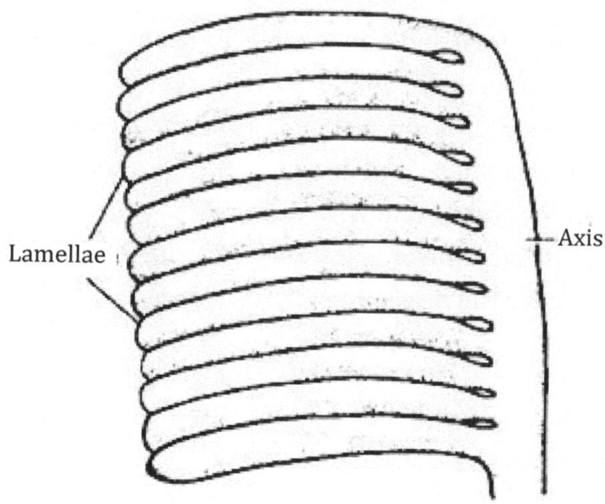

Figure 2.4: Pecten of Fowl.

Structure of Pecten of Fowl

1. It is present in reptiles and birds.
2. It is absent in crocodile, alligator and kiwi bird.
3. Pecten is membranous, folded and somewhat rectangular in shape.
4. It projects into the cavity of eyeball.
5. It is well developed in diurnal bird.
6. It is black pigmented and longitudinal fan like structure.
7. It is highly vascular.
8. Structurally, it shows axis and many lamellae with serrate margins.

Function

a. It helps in accommodation, focussing and perception of movements of objects.
b. It brings oxygen and nourishment to the retina, which in birds has no capillary circulation.
c. It has taxonomic importance (Ghose and Manna, 2016).

Columella of Fowl

Aim: Temporary preparation of columella of fowl.

Materials

Head of fowl, Alizarin red stain and glycerine.

Procedure

1. Take a head of fowl and remove the ear drum from anterior side.
2. Locate columella in the auditory cavity.
3. Put a drop of glycerine on columella location for easy and quick separation.
4. Remove it with the help of forcep.
5. Kept in watch glass, stain with Alizarin red
6. Mount with glycerine or water on glass slide.
7. Observe under the dissecting microscope.

Location

Columella lies in the cavity of middle ear (tympanic cavity).

Structure of Columella of Fowl

1. Columella is bony rod like structure.
2. It has stapedial plate (stape) at one end and tympanic membrane at other end.

3. Stape is the smallest bone in the vertebrates.
4. Tympanic membrane is also called ear drum.
5. It has supra-stapedial, infra-stapedial and extra-columellar cartilages.
6. Stape is the smallest bone plate which plugs into the fenestra ovalis.
7. Fenestra ovalis is small opening of internal ear.

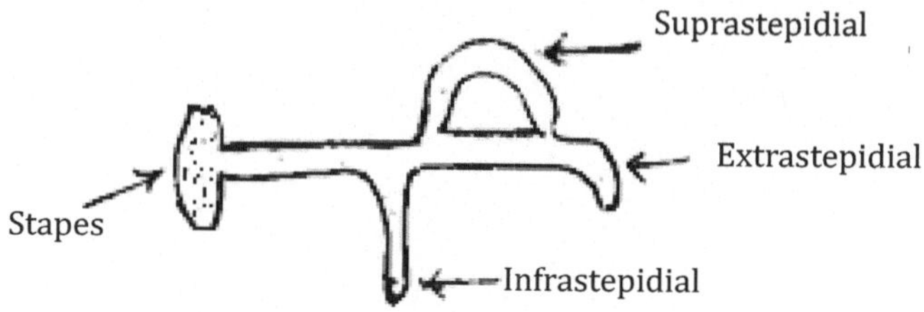

Figure 2.5: Columella of Fowl.

Function

1. It helps in hearing by transmitting sound waves towards internal ear.

Chapter

3

Scales of Fishes

Cycloid Scale of *Labeo*

Aim: Temporary preparation of cycloid scale of *Labeo.*

Materials

Bony fish like *Labeo,* Alizarin red stain and glycerine.

Procedure

1. Remove a cycloid scale of *Labeo*.
2. Put scale in a watch glass containing 10 per cent KOH solution.
3. Stir slowly with a needle till the covering epithelium dissolves.
4. Wash thoroughly with water to remove the trace of KOH.
5. Take cycloid scale in a watch glass.
6. Add 2 drop of Alizarin red stain in a watch glass and stain for 2 minutes.
7. Remove excess stain and mount with glycerine.
8. Put cover slip, blot with blotting paper and observe under dissecting microscope.

Location

Cycloid scales are present on the skin of bony fishes like *Labeo*.

Structure of Cycloid Scale

1. Scales are derivatives of mesenchymal cells of dermis.
2. Some fishes are **"naked"** devoid of scales, *e.g.*, freshwater catfish.

3. Cycloid scales are visible, thin, plate of bone with a semicircular free border.
4. Presence of cycloid scales is characteristics of burbot fish (an elongated bottom dwelling fish that is the only freshwater member of the cod family) and softy-rayed bony fishes.
5. Each cycloid scale has anterior embedded part and posterior free part.
6. Cycloid scale is thin at the periphery and thick at the centre.
7. Nucleus, radius and lines of growth are present.
8. Scale hasa several concentric lines of growth which can be used for determining the age of fish.
9. The cycloid scales are devoid of teeth or spines, hence seem cyclic.
10. In Micropterus fish, both cycloid and ctenoid scales are found.

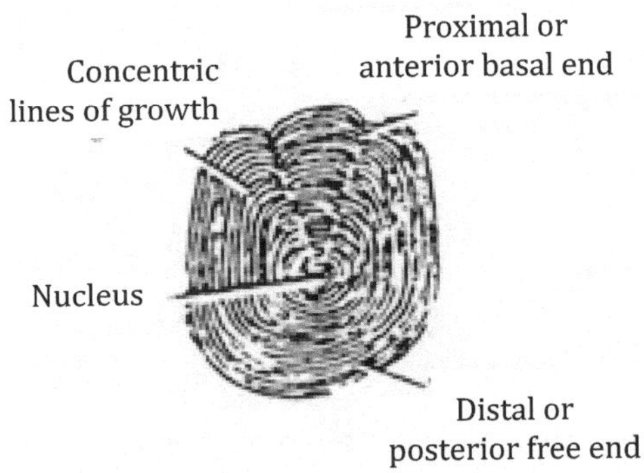

Figure 3.1: Cycloid Scale of *Labeo.*

Function

1. It provides exoskeleton and protection to the body.
2. It has taxonomic importance. It provides flexibility.
3. It minimizes friction during swimming.
4. Lines of growth used for determination of age of fish.

Ctenoid Scale of Flat Fish

Aim: Temporary preparation of ctenoid scale of flat fish (lepa or sole fish).

Materials

Ctenoid scales are found bony fish like flat lepa fish, *Cynoglossu cynoglossus.* Lea fish is also called sole fish. It is also found in Koi fish, *Anabas testudineus.* Alizarin red stain and glycerine.

Procedure

1. Remove one ctenoid scale of flat fish (lepa or sole fish).

2. Put scale in a watch glass containing 10 per cent KOH solution.
3. Stir slowly with a needle till the covering epithelium dissolves.
4. Wash thoroughly with water to remove the last trace of KOH.
5. Take ctenoid scale in a watch glass.
6. Add 2 drop of Alizarin red stain in a watch glass and stain for 2 minutes.
7. Remove excess stain and mount with glycerine.
8. Put cover slip, blot with blotting paper and observe under dissecting microscope.

Location

Ctenoid scales are present on the skin of bony fishes like flat fish (lepa fish).

Structure Ctenoid Scales

1. Ctenoid scales are found in bony fish like flat fish, *Cynoglossu cynoglossus.*
2. Ctenoid scale is the characteristics of spiny-rayed bony fishes (Acanthopterygii).
3. Flat fish has tongue like body. Both eyes present on right or left side.
4. Ctenoid scales are also found in koi fish, *Anabas testudineus.*
5. Ctenoid scales have characteristic teeth at its posterior part.
6. Nucleus, radius and lines of growth are present.
7. Ctenoid scale has anterior embedded part and posterior or distal free part.
8. They are arranged obliquely in such a manner that the posterior end of one scale overlaps the anterior edge of the scale present behind.

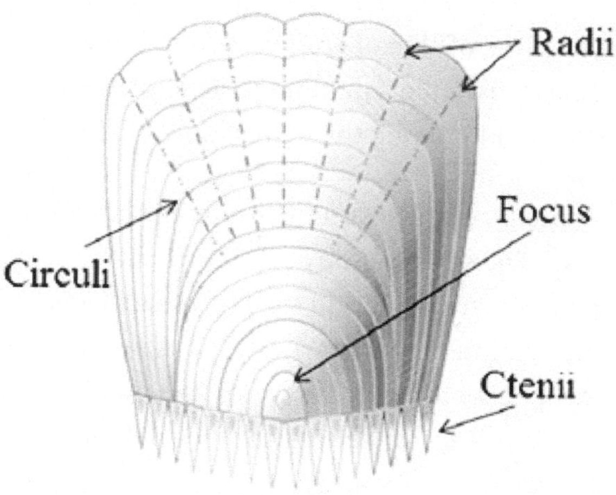

Figure 3.2: Ctenoid Scale of Lepa Fish.

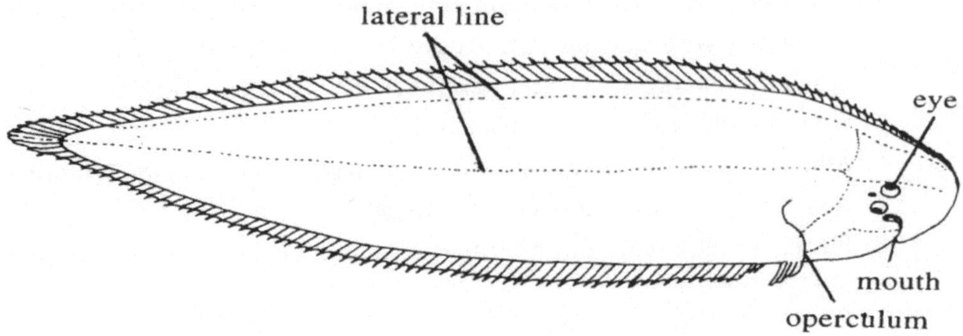

Figure 3.3: Flat Lepa Fish, *Cynoglossu cynoglossus.*

Function

1. It provides exoskeleton and protection to the body.
2. It has taxonomic importance. It provides flexibility.
3. It minimizes friction during swimming.
4. Lines of growth used for determination of age of fishes (Ghose and Manna, 2016).

Placoid Scale of *Scoliodon*

Aim: Temporary preparation of placoid scale of *Scoliodon*.

Materials

Cartilage fish like *Scoliodon*, Alizarin red stain and glycerine.

Procedure

1. Take a small piece of skin of *Scoliodon* from dorsal surface.
2. Put it in a glass test tube containing 10 per cent KOH solution.
3. Boil with constant stirring the help of spirit lamp till the skin dissolves completely.
4. Pour the contents of the test tube in a large watch glass.
5. Allow to cool. The scales settle at the bottom.
6. Decand the fluid with water till the last trace of KOH is removed.
7. Take 2 drop of suitable stain in a watch glass and stain for 2 minutes.
8. Pipette a drop of water with the scales and put it on slide.
9. Remove water and excess stain by a piece of blotting paper.
10. Mount with glycerine or water.
11. Put cover slip and observe under compound microscope (Mutakekar *et al.*, 2013, Ghose and Manna, 2016).

Location

Placoid scales are present on the skin of cartilage fishes like *Scoliodon.*

Structure of Placoid Scale

1. Placoid scales are microscopic in nature.
2. Each placoid scale has a base and a body.
3. Body of placoid scale is formed by trident spine and base of placoid scale is formed by diamond shaped basal plate which is embedded in the skin.
4. Trident spine is formed by two lateral and one median spines.
5. Basal plate is and embedded in the skin.
6. The central cells do not calcify and constitute the pulp.
7. It is formed by calcified tissue and the cement substances.
8. The mesenchymal cells of the dermis secrete the basal plate.
9. Basal plate encloses a cavity called pulp cavity which is filled with pulp.
10. Pulp contains odontoblasts which secrete dentine.
11. Dentine is covered by a hard layer of vitrodentine.
12. Pulp contains connective tissue, blood vessels and nerve fibres.
13. The spines of placoid scale erupt from the cells of epidermis and projects out, while the basal plate lies embedded in the dermis.

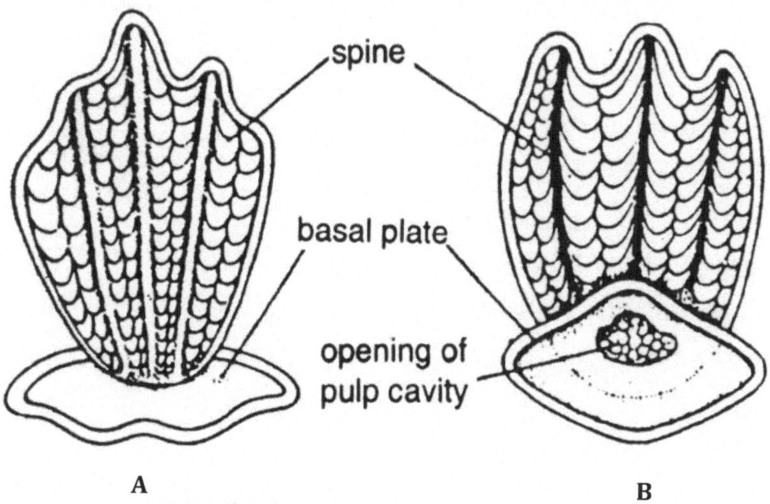

Figure 3.4: Placoid Scale of *Scoliodon.*
A: Dorsal view; B: Ventral view.

Function

1. It provides exoskeleton and protection to the body.
2. Dried skin is used as a polish paper for polishing the furnitures and metals.

3. It has taxonomic importance.
4. Scales are not found in the lampreys and hagfishes.
5. Sharks are characterized by presence of placoid scales.
6. Primitive bony fishes possess ganoid scales.
7. The higher bony fishes have ctenoid or cycloid scales.
8. Fish scale count is highly important in taxonomy.
9. The number of scales present in lateral line, along and around the body, is specific in every species.
10. Fish age can be determined by measuring space in annual rings of the scales.
11. Species like Atlantic Salmon, scales exhibit the presence of spawning marks.

Chapter

4

Biochemical Tests

Aim

Qualitative tests to identify functional groups of carbohydrates and lipid in given solutions: Lipid, Sucrose, Glucose, Fructose and Lactose.

Requirements

Test tubes, test tube holders, boiling water bath, spirit lamp, thermometer, glass beakers, glass slides, compound microscope and chemicals.

Table 4.1: Preliminary Tests

Test	*Observation*	*Inference*
Preliminary tests. **Solubility test:** 1 mL. O. S. + 1 mL of Conc. H_2So_4.	Substances soluble with or without change in colour.	Aldehydes, ketones or esters containing substances may be present.
	Substances insoluble	Hydrocarbons (Alkanes) containing substances may be present.
Test for esters: 1 mL. O. S. + 2 drops of phenolphthalein + 2 drops of 10 per cent NaOH heat on lamp.	Pink color slowly disappears.	Ester containing substances O=C-O group) may be present.
	No pink colour	Aldehydes containing substances (-CHO group) may be present.
Test for aldehydes: 1 mL. O. S. + 2 drops of Schiff's reagent, shake well.	Pink colour	Aldehydes containing substances (-CHO group) may be present.

Test	*Observation*	*Inference*
Benedict's test: 1 mL O. S. + 2 drops of Benedict's reagent. Mix thoroughly and heat for 2 minutes and cool under tap water.	No yellow ppt	Sucrose may be present (Perform test for sucrose)
	Yellow ppt of cuprous oxide appears (Yellow, green, red, orange and brick color depends on concentration of sugar.)	Reducing glucose, fructose and lactose may be present (Test for glucose, fructose and lactose.)

Table 4.2: Test for Lipid

Test	*Observation*	*Inference*
Grease spot test: Take 2 drops of O.S. on ordinary writing paper.	Paper becomes translucent and greasy.	Lipid present
Emulsification test: Take 1 mL of O.S. in beaker add 100 mL tap water.	Lipid droplets are observed.	Lipid present
1) Solubility test: 1.1) 1mL O. S. + 1mL of Chloroform.	Lipid gets soluble.	Lipid present
1.2) 1mL O. S. + 1mL of absolute alcohol.	Lipid gets settle at the bottom of test tube.	Lipid present
2) Sudan III test: Take 0.5 mL chloroform in test tube and add 1 mL O. S. drop by drop till the sample is fully dissolved. Add one drop of Sudan III reagent.	Red colour appeared	Lipid present
3) Saponification test: 1mL O. S. in test tube + 20 drops of 40 per cent NaOH + 2 drops of glycerol. Boil gently for 3 minutes. If oil globules are visible boiling must be continued. Cool and divide the solution in 3 equal parts in test tube 1, 2 and 3.	Soap is formed and it rises to the surface	Oil confirmed
3.1) In test tube 1 add 5 drops of Conc. HCl.	Colourless layer of fatty acids separates out and floats to the surface.	Lipid is confirmed
3.2) In test tube 2 add crystals of NaCl.	White ppt soap separates out and floats to the surface.	Lipid is confirmed
3.3) In test tube 3 add 10 drops of 10 per cent $CaCl_2$.	Colourless layer of fatty acids separates out and floats to the surface.	Lipid is confirmed

Result

The given unknown solution is Lipid containing ester group.

Table 4.3: Test for Sucrose

Test	*Observation*	*Inference*
1) Seliwanoff's test: 1 mL. O. S. + 5 drops of Con. HCl + 5 drops of Seliwanoff's reagent heat for 2 minutes if necessary cool under tap water.	Cherry red to orange color	Sucrose present and confirmed.
2) Methylene blue test: 1 mL. O. S. + 2 drops of dilute Methylene blue + 1 mL 10 per cent NaOH heat for 2 minutes and cool.	Blue color of the solution disappears (methylene blue reduced to colourless leucomethylene blue	Sucrose present and confirmed.
3) Inversion test: 1mL. O. S. + 2 drops of Conc. HCl and boil for 2 minutes cool, Make alkaline with 4 drops of 10 per cent NaOH. Add Benedict's reagent and heat for 2 minutes.	Yellow or Brick red ppt	Sucrose present and confirmed.

Result

The given unknown solution is sucrose containing ketone group.

Table 4.4: Test for Glucose, Fructose and Lactose

Test	*Observation*	*Inference*
Molisch's test: 1 mL. O. S. + 2 drops of Molisch's reagent and mix thoroughly. Incline test tube and add 1 mL Conc. H_2So_4 so that do not mix.	A purple or pink ring appears at the junction of the two solutions.	Carbohydrates like glucose, fructose and lactose may be present.
Osazone test for glucose, fructose and lactose: 1 mL O. S. + 8 drops of glacial acetic acid + few amount of Phenyl hydrazine hydrochloride powder+ twice amount of sodium acetate crystals. Keep the test tube in boiling water bath. Take osazone material on glass slide and observe under compound microscope.	Fan shaped (needle like) yellow coloured crystals of glucosazone or fructosazoe are observed within 15 minutes.	Glucose or fructose is present.
Seliwanoff's Test for separation of glucose and fructose: 1 mL O. S. + 5 drops of Seliwanoff's reagent heat for 1 minute.	No Cherry red to orange color.	Glucose present and confirmed.
	Cherry red to orange color	Fructose present and confirmed.

Test	*Observation*	*Inference*
Test for ketones:1 mL. O. S. + 2 drops of sodium nitroprusside solution + 2 drops of 10 per cent NaOH.	Red colour	Fructose containing substance(C=O group) present and confirmed.
	Thin yellow prickle ball or cotton ball shaped crystals of fructosazoe are observed within 25 minutes.	Lactose present and confirmed.

Result

The given unknown solution is glucose containing aldehyde group.

The given unknown solution is fructose containing ketone group.

The given unknown solution is lactose containing aldehyde group.

Chapter 5

Activity of Salivary Amylase

Study of Activity of Salivary Amylase under Optimum Conditions

Aim

Study of activity of salivary amylase under optimum conditions.

Principle

Leuchs (1831) described the hydrolysis of starch by salivary amylase. Salivary amylase catalyses the breakdown of starch into maltose and glucose which can be confirmed by Bendict's reagent. The amylase is active at pH 6.8, 1 per cent starch, temperature 37 °C and 15 minutes of time.

Requirements

1. **Own saliva:** Clean mouth with drinking water. Hold a piece of sour foodstuff like tamarind piece or such material in front of the tongue. Collect saliva from under the tongue by clean dropper.
2. **1 per cent starch solution:** Take 1 g of starch powder in 5 mL distilled water and mix thoroughly. Add 95 mL distilled water and heat till boiling point with stirring constantly. Cool and add a few drops of toluene or chloroform as a preservative.
3. **I_2 solution:** Take 1 g of iodine and 1 g of potassium iodine in 100 mL of water.

4. **pH 6.8 :** It is prepared from tris maleate buffer or phosphate buffer or any other suitable buffer: Prepare pH 6.8 solution from tris maleate buffer by using solution 'A' and solution 'B'.

   Solution (A): 0.2 M solution of Tris acid maleate (24.2 g of tris (hydroxymethyl) aminomethane + 23.2 g of maleic acid or 19.6 g of maleic anhydride in 1000 mL of distilled water).

   Solution (B): 0.2 M NaOH in 1000 mL of distilled water. 50 mL of solution 'A' + 'X' mL of solution 'B' diluted to a total of 200 mL in distilled water.

Table 5.1: Preparation of Tris Maleate Buffer pH 6.8

'A' Solution in mL	*'X' mL of 'B' Soloution*	*Dilution to 200 mL Dist. Water*	*pH*
50	45.0	105	6.8

5. **Benedict's reagent.**

Table 5.2: Observations of Salivary Amylase Activity under Optimum Condition

Test Tube No.	*Test*	*Observation*	*Inference*
1.	**Starch test:** 2 mL starch solution + 2 drops of I_2 solution.	Blue colour suspension	Starch present
2.	**Starch-amylase incubation:** 1 mL of pH 6.8 solution + 2 mL starch solution + 1 mL salivary amylase. Keep for incubation in water bath at 37 ° C for 20 minutes (Perform starch digestion test and Benedict's test.		
3.	**Starch digestion test:** 2 mL of incubated solution + 2 drops of I_2 solution.	No blue colour	Starch digested
	Benedict's test: 1 mL of incubation solution + 2 drops of Benedict's reagent. Mix thoroughly and heat for 2 minutes and cool under tap water.	Yellow ppt.	Reducing maltose and glucose present.

Result

Salivary amylase catalyses starch to forms maltose and glucose under optimum condition at pH 6.8, temperature 37 °C, time 15 minutes and 1 per cent starch concentration (Leuchs, 1831, Gomori 1955, Ghose and Manna, 2016).

Effect of Temperature, pH and Salinity on Activity of Salivary Amylase

Aim

Effect of temperature, pH and salinity on activity of salivary amylase

Principle

Leuchs (1831) described the hydrolysis of starch by salivary amylase. Salivary amylase catalyses the breakdown of starch into maltose and glucose which can be confirmed by Bendict's reagent. The amylase is active at pH 6.8, 1 per cent starch, temperature 37 °C and 20 minutes of time.

Requirements

1. **Own saliva:** Clean mouth with drinking water. Hold a piece of sour foodstuff like tamarind piece or such material in front of the tongue. Collect saliva from under the tongue by clean dropper.
2. **1 per cent starch solution:** Take 1 g of starch powder in 5 mL distilled water and mix thoroughly. Add 95 mL distilled water and heat till boiling point with stirring constantly. Cool and add a few drops of toluene or chloroform as a preservative.
3. **pH 6.8, pH 8.6 and pH 5.2 solutions:** Prepare pH 6.8, pH 8.6 and pH 5.2 from tris maleate buffer by using solution 'A' and solution 'B'.

   Solution (A): 0.2 M solution of Tris acid maleate (24.2 g of tris (hydroxymethyl) aminomethane+ 23.2 g of maleic acid or 19.6 g of maleic anhydride in 1000 mL of distilled water).

   Solution (B): 0.2 M NaOH in 1000 mL of distilled water. 50 mL of solution A + 'X' mL of solution B diluted to a total of 200 mL in dist. water.

Table 5.3: Preparation of Tris Maleate Buffer pH 5.2, pH 6.8 and pH 8.6

'A' Solution in mL	*'X' mL of 'B' Soloution*	*Dilution to 200 mL Dist. Water*	*pH*
50	7.0	143	5.2
50	45.0	105	6.8
50	86.5	63.5	8.6

4. **I_2 solution:** Take 1 g of iodine and 1 g of potassium iodine in 100 mL of water.
5. **2 per cent, 0.9 per cent and 0.5 per cent NaCl solutions:** Prepare 2 per cent, 0.9 per cent and 0.5 per cent of NaCl solutions by adding 2 grams, 900 mg and 500 mg of NaCl crystals in 100 mL of distilled water separately.

6. **37 °C, 60 °C and 4 °C water:** They take in separate glass beakers for enzyme incubation.

Table 5.4: SET 1: Effect of pH on Amylase Activity

Test Tube No.	*pH 1 mL*	*1 per cent Starch in mL*	*Salivain mL*	*Temp. °C*	*Time in Min.*	*I_2 in Drops*	*Obs.*	*Inferences*
1.	6.8	2	1	37	15	2	No blue colour	Maltose and glucose present
2.	8.6	2	1	37	15	2	Blue colour	Starch present
3.	5.2	2	1	37	15	2	Blue colour	Starch present

Table 5.5 SET 2: Effect of Temperature on Amylase Activity

Test Tube No.	*pH 6.8 in mL*	*1 per cent Starch in mL*	*Salivain mL*	*Temp. °C*	*Time in Min.*	*I_2 in Drops*	*Obs.*	*Inferences*
1.	1	2	1	37	15	2	No blue colour	Maltose, glucose present
2.	1	2	1	60	15	2	Blue colour	Starch present
3.	1	2	1	4	15	2	Blue colour	Starch present

Table 5.6 SET 3: Effect of Salinity on Amylase Activity

Test Tube No.	*NaCl 1 mL*	*1 per cent Starch in mL*	*Salivain mL*	*Temp. °C*	*Time in Min.*	*I_2 in Drops*	*Obs.*	*Inferences*
1.	0.9 per cent	2	1	37	15	2	No blue colour	Maltose and glucose present
2.	2 per cent	2	1	37	15	2	Blue colour	Starch present
3.	0.5 per cent	2	1	37	15	2	Blue colour	Starch present

Result

SET.1: The given enzymatic reaction is effect of pH on salivary amylase activity. At pH 6.8 amylase enzyme acts on starch and produces maltose and glucose. At pH 8.6 and pH 5.2 amylase enzyme does not act on starch and starch gives blue colour to the I_2 solution. At pH 5.2 and pH 8.6 enzyme is inhibited.

SET.2: The given enzymatic reaction is effect of temperature on salivary amylase activity. At temperature 37 °C amylase enzyme acts on starch and produces maltose and glucose. At temperature 60 °C and 4 °C amylase enzyme does not act on starch and starch gives blue colour to the I_2 solution. At temperature 60 °C enzyme is denatured and at 4°C enzyme is inhibited.

SET.3: The given enzymatic reaction is effect of salinity on amylase activity. At 0.9 per cent salinity amylase enzyme acts on starch and produces maltose and glucose. At 2 per cent and 0.5 salinity, amylase enzyme does not act on starch and starch gives blue colour to the I_2 solution. At 2 per cent and 0.5 salinity, amylase enzyme is inhibited (Leuchs, 1831, Gomori 1955 and Ghose and Manna, 2016).

Chapter 6

Estimation of Total Protein and Lipid

Estimation of Total Protein from given Solution by Lowry's Method

Aim

To estimate total protein from given solution by Lowry's method.

Principle

Cuprous ion reduces the Folin-Ciocalteu reagent. The phenolic group of tyrosine in protein produces blue purple colour. The optical density is measured by calorimeter at 670 nm. The intensity of color is proportional to the protein concentration (Lowry *et al.*, 1951, Ghose and Manna, 2016).

Requirements

1. **1 per cent insect larval homogenate:** The insect larvae were immobilised by chilling on ice for 15minutes. The larvae were weighed and homogenised in the cold double distilled water using a ground glass mortar and pestle. The homogenate was diluted with cold double distilled water so as to get 1 per cent concentration. The homogenate was taken in tube and centrifuged at 3000 rpm for 30 minutes and the supernatant obtained after centrifugation was used for protein estimation.
2. **Reagent A: Sodium carbonate 2 per cent:** 20 gm of Na_2CO_3 in 1 litre 0.1 N NaOH. For 0.1N NaOH take 4 gm NaOH in 1000 mL of distilled water.
3. **Reagent B: Sodium-potassium tartrate 1 per cent:** 10 gm sodium-potassium tartrate 1 litre distilled water.
4. **Reagent C: Copper sulphate 0.5 per cent:** 5 gm $CuSO_{4,}5H_2O$ in 1 litre distilled water.

5. **Reagent D (I):** Mix 48 mL reagent A + 1 mL reagent B + 1mL reagent C.
6. **Reagent E (II):** Dilute Folin and Ciocateu's reagent with an equal volume of distilled water just before use.
7. **Preparation of standard curve (BSA, bovine serum albumin, 1mg/ mL):**
   7.1. To determine protein values, firstly standard protein graph was drawn by taking 0.2 per cent BSA in distilled water (Ghose and Manna, 2016) and kept in freeze over night for obtaining the clear solution. Take 200 gm of BSA in 100 mL of distilled water for 0.2 per cent BSA.
   7.2. Add 0.1 to 0.9 mL of BSA in test tubes making total volume of 1 mL by distilled water. Stock standard solution concentration was accorded as 1 mg/mL by using distilled water (Ghose and Manna, 2016).
   7.3. Take 0.5 mL solution in separate test tubes.
   7.4. Add 4.5 mL of reagent I in each test tube and incubate for 10 minutes.
   7.5. Add 0.5 mL of reagent II in each test tube and incubate for 30 minutes. Take sufficient amount in glass cuvette.
   7.6. Measure optical density of different BSA solutions of known concentrations. Record the colorimeter readings at 670 nm. Plot the concentration of BSA on X-axis and optical density on Y-axis on a graph paper. Draw a line touching the points on the graph paper; it will be a straight line.
8. **Reagent blank:** Put 1 mL distilled water, 4.5 mL of alkaline solution of Reagent D (I) and 0.5 mL of diluted solution of Reagent D (I). Mix thoroughly and switch calorimeter, wait for 5 minutes for warming up. Fill a cuvette with reagent blank and adjust the optical density to zero after setting the 670 nm in colorimeter (Ghose and Manna, 2016).

Table 6.1: Data sheet for standard graph for bovine serum albumin.

Sl. No.	*BSA (mL)*	*BSA in μg/mL*	*Conc. of BSA in mg/mL*	*Dist. Water in mL*	*Reagent I (mL)*		*Reagent II (mL)*		*Optical Density (O.D.)*
1	0.1	10	0.010	0.9	4.5	***Room Temperature Incubate 10 min.***	0.5	***Room Temperature Incubate 30 min.***	0.1 7
2	0.2	20	0.020	0.8	4.5		0.5		0.25
3	0.3	30	0.030	0.7	4.5		0.5		0.34
4	0.4	40	0.040	0.6	4.5		0.5		0.43
5	0.5	50	0.050	0.5	4.5		0.5		0.51
6	0.6	60	0.060	0.4	4.5		0.5		0.59
7	0.7	70	0.070	0.3	4.5		0.5		0.69
8	0.8	80	0.080	0.2	4.5		0.5		0.76
9	0.9	90	0.090	0.1	4.5		0.5		0.85
Blank	-	-	-	1.0	4.5		0.5		Adjust zero by blank

9. **Estimation of total protein from insect larvae.**
   - 9.1. Take 0.5 mL 1 per cent insect larval homogenate in test tube.
   - 9.2. Add 4.5 mL of reagent I, mixed well and incubate at room temperature for 10 minutes.
   - 9.3. Add 0.5 mL reagent II, mixed well and incubate at room temperature for 30 minutes.
   - 9.4. The blue purple color is obtained. The optical density is measured by calorimetrically at 670 nm.
   - 9.5. Standard graph of protein is used for estimation of protein (Lowry *et al.,* 1951, Ghose and Manna, 2016).

Result

The estimated total protein from insect larvae is -- mg/mL.

Estimation of Total Lipid

Aim

To estimate total lipid from given solution.

Principle

Lipid reacts with sulphuric acid to form carbonium ions which react with vanillin-phosphoric acid to yield a purple color. The optical density is measured by calorimeter at 540 nm. The intensity of color is proportional to the lipid concentration.

Requirements

1. **1 per cent insect larval homogenate in distilled water.**
2. **Vanillin-phosphoric acid reagent (Sulfo-phospho-vanillin method):** 6 mg of Vanillin (98 per cent) was dissolved in 100 mL of hot distilled water and stirred with 500 mL of 85 per cent ortho-phosphoric acid and kept in the dark.
3. **Conc. H_2SO_4.**
4. **Mixture of chloroform and methanol:** Mix 10 mL chloroform in 20 mL of methanol in the ratio of 1: 2.
5. **Standard graph of lipid (Olive oil, 1mg/mL):**
   - 5.1. To determine lipid values, firstly standard lipid graph was drawn by taking 0.1 per cent olive oil emulsion in distilled water. Take 100gm of olive oil in 100 mL of distilled water for 0.1 per cent olive oil (Bozdogan *et al.,* 2016).
   - 5.2. Add 0.1 to 0.9 mL of olive oil emulsion test tubes making total volume of 1 mL by mixture of chloroform and methanol. Stock standard solution concentration was accorded as 1 mg/mL by using mixture of chloroform and methanol (1:2).

5.3. Take 0.2 mL of these solutions in test tubes. Heat in water bath at 100 °C until chloroform and methanol mixture vaporise completely.

5.4. Add 0.18 mL Conc. H_2SO_4, vortex tubes and heat in bath at 100 °C for 2 min.

5.5. Add 0.5 mL vanillin-phosphoric acid and incubate at 37 °C for 15 minutes.

5.6. Cool keeping in dark box for 45 minutes and purple color appeared.

5.7. Take optical density of different olive oil solutions of known concentrations.

5.8. Record the colorimeter readings at 540 nm. Plot a graph of concentrations of olive oil on X-axis and optical densities on Y-axis. Draw a line touching the points on the graph paper; it will be a straight line.

6. **Reagent blank:** Put 1 mL distilled water. Add 0.18 mL Conc. H_2SO_4 and mixed by vortex and heated in the bath at 100 °C for 2 minutes. Add 0.5 mL reactive vanillin-phosphoric acid incubates at 37 °C for 15 minutes and for purple color formation cool keeping in dark box for 45 minutes. Switch calorimeter; wait for 5 minutes for warming up. Adjust the optical density to zero by blank after setting the 540 nm in colorimeter (Bozdogan *et al.*, 2016 and Anschau *et al.*, 2017).

Table 6.2: Data Sheet for Preparation of Standard Graph for Lipid (Olive oil)

Sl.No.	*Olive Oil (mL)*	*Conc. of Olive Oil µg/mL*	*Conc. of Olive Oil mg/mL*	*Chloroform and Methanol in mL*	*Conc. H_2SO_4 (mL)*	**Heat in Water Bath at 100 °C for 2 min.**	*Vanallin-phosphoric Acid in mL*	**Mix and Incubate 15 min. at 37 °C**	*Optical Density (O.D.)*
1	0.1	10	0.010	0.9	0.18		0.5		0.21
2	0.2	20	0.020	0.8	0.18		0.5		0.30
3	0.3	30	0.030	0.7	0.18		0.5		0.21
4	0.4	40	0.040	0.6	0.18		0.5		0.53
5	0.5	50	0.050	0.5	0.18		0.5		0.64
6	0.6	60	0.060	0.4	0.18		0.5		0.75
7	0.7	70	0.070	0.3	0.18		0.5		0.80
8	0.8	80	0.080	0.2	0.18		0.5		0.88
9	0.9	90	0.090	0.1	0.18		0.5		0.99
Blank	-	-	-	1.0	0.18		0.5		Adjust zero

7. **Estimation of total lipid from insect larvae.**

7.1. Take 0.2 mL of of 1 per cent insect larval homogenate in test tube.

7.2. Add 1 mL mixture of chloroform and methanol and kept in water bath at 100 °C until complete evaporation.

7.3. Add 0.18 mL Conc. H_2SO_4, vortex tubes and heat in water bath at 100 °C for 2 minutes.

7.4. Add 0.5 mL vanillin-phosphoric acid and incubate at 37 °C for 15 minutes.

7.5. Cool keeping in dark box for 45 minutes and purple color appeared.

7.6. Take optical density by colorimeter at 540 nm. Use standard graph of olive oil for estimation of lipid (Bozdogan *et al.*, 2016 and Anschau *et al.*, 2017).

Result

The estimated total lipid from insect larvae is ---mg/mL.

Chapter 7

DNA Isolation and Estimation of Uric Acid

DNA Isolation from Cauliflower

Aim: To isolate DNA from cauliflower.

Principle

DNA is the genetic material. The extraction of DNA includes rupturing of tissues, cell membranes, nuclear membranes followed by deproteinization and finally prtecipitation of the nucleic acids using ethanol. The positive ions of sodium chloride and sodium citrate bind with negatively charged DNA backbone, creating a DNA molecule with neutral charge. Chilled ethanol removes bound water from DNA and results precipitation or aggregation of DNA (Ghose and Manna, 2016).

Requirements

1. 10 grams of cauliflower.
2. Saline sodium citrate (Homogenization medium 1 L): It included 500 mL 0.15 M NaCl + 500 mL 0.2 N sodium citrate. Sodium chloride (1M) is prepared by addition of 58.44 gram of NaCl in 1L and 0.15 M NaCl is prepared by addition of 8.766 gram of NaCl in 1L. Sodium citrate (1M) is prepared by addition of 29.41 gram of sodium citrate in 1L and 0.2 M sodium citrate is prepared by addition of 5.882 gram of sodium citrate in 1L.
3. Cold ethanol 1 L: Use chilled ethanol.

4. Mortar and pestle, cutting board or knife, muslin cloth, glass rods, pipettes, beaker, sprit lamp, centrifuge tubes, centrifuge machine and freeze.

Procedure

1. Take 10 g of cauliflower and make small pieces by cutting board or knife.
2. Take 10 g of crushed cauliflower and homogenize with 100 mL of saline sodium citrate in mortar and pestle and heat under spirit lamp for 5 minutes.
3. The pureed (disrupted or homogenized) cauliflower is filtered with muslin cloth to remove any solid material resulting in a clear cell homogenate.
4. Centrifuge at 6000 rpm for 10 minutes.
5. Take supernatant and discard residue.
6. Supernatant again centrifuge at 6000 rpm for 10 minutes.
7. Take supernatant and add ice cold ethanol by inclining the test tube.
8. Without mixing observe white precipitation at surface layer. DNA aggregation found the upper layer.

Take 5 g of cauliflower and make small pieces by cutting board.

⇩

Homogenize with 100 mL of saline sodium citrate and heat under sprit lamp for 5 minutes.

⇩

Filtered with muslin cloth.

⇩

Centrifuge at 6000 rpm for 10 minutes.

⇩

Take supernatant and discard residue.

⇩

Again centrifuge supernatant at 6000 rpm for 10 minutes

⇩

Add 1 mL of ice cold ethanol in inclined tube slowly from the margin.

⇩

DNA aggregates in white precipitation at the upper layer

Figure 7.1: Flow Chart for DNA Isolation from Cauliflower.

Result

DNA aggregation in white precipitation at the upper layer of test tube is observed.

Estimation of Uric Acid from Poultry Bird Excreta

Aim: To estimate uric acid from poultry bird excreta.

Principle

Uric acid is the end product of purine metabolism. Hydrochloric acid and ammonium chloride used in the separation of uric acid. Uric acid is precipitated by ammonium sulfate. Filtrate containing uric acid was kept in Goach crucible and after suction under suction pump and drying the weight of uric acid was estimated.

Requirements

1 g of poultry bird excreta, 0.15 N NaCl, 20 per cent HCl, 0.25 N Piperidine, phenolphthalein, ammonium chloride, 10 per cent ammonium sulphate, Goach crucible, suction pump, whatman's filter paper, muslin cloth, glass rods, pipettes, beaker, freeze, 100 mL volumetric flask, water bath and oven, 1 g of poultry bird excreta.

Procedure

1. Take 1g of poultry bird excreta in 100 mL volumetric flask.
2. Add 10 mL 0.15 N HCl and boil for 3 minutes and cool.
3. Add 5 mL 0.25 N Piperidine for making alkaline and confirm by 2 drops of phenolphthalein. Boil for 45 minutes at 60 °C and cool.
4. Make to mark of 100 mL volumetric flask and filter by whatman's filter paper.
5. Take 50 mL filtrate and add 16.5 g of ammonium chloride. Boil for 20 minutes at 60 °C and cool. Boil for 20 minutes at 60 °C and cool.
6. Keep for crystallization for 24 hours and filter by whatman's filter paper.
7. Precipitate was washed with 10 per cent ammonium sulphate and with distilled water.
8. Collect in beaker and add 5 mL 20 per cent HCl.
9. Boil for 15 minutes at 60 °C making 15 mL volume.
10. Keep for crystallization for 24 hours and filter into weighed Goach crucible. Dried at 100 at °C and weighed the amount of uric acid (John and Johnson, 1931).

Take 1g of poultry bird excreta in 100 mL volumetric flask.

⇩

Add 10 mL 0.15 N HCl and boil for 3 minutes and cool.

⇩

Add 5 mL l 0.25 N Piperidine and confirm by 2 drops of phenolphthalein.

⇩

Boil for 45 minutes at 60 °C and cool.

⇩

Make to mark of 100 mL volumetric flask and filter by whatman's filter paper.

⇩

Take 50 mL filtrate and add 16.5 g of ammonium chloride.

⇩

Boil for 20 minutes at 60 °C and cool.

⇩

Keep for crystallization for 24 hours and filter by whatman's filter paper.

⇩

Precipitate was washed with 10 per cent ammonium sulphate and with distilled water.

⇩

Add 5 mL 20 per cent HCl, Boil for 15 minutes at 60 °C make 15 mL volume.

⇩

Boil for 15 minutes at 60 °C making 15 mL volume.

⇩

Keep for crystallization for 24 hours and filter into crucible and suck by pump.

⇩

Dried at 100 at °C and weighed the amount of uric acid.

Figure 7.2: Flow Chart for Estimation of Uric Acid.

Observation

Wt. of crucible before drying –g and wt. of crucible after drying––g.

Result

The estimated uric acid from poultry bird excreta is ---- g.

Significance

Increase in uric acid may be due to renal failure and other diseases.

Chapter

8

Economic Importance of Insects

Identifying Features and Economic Imporance of Insects

1. *Helicoverpa armigera* (Hubner)

1.1. Systematic Position of *H. armigera*

Kingdom : Animalia

Phylum : Arthropoda

Class : Insecta

Subclass : Pterygota

Division : Endopeterygota

Order : Lepidoptera

Family : Noctuidae

Genus : *Helicoverpa* (*Heliothis*)

Species : *armigera* (Hubner)

Common name: Army worm, gram pod borer, American ball worm, corn ear worm, cotton ball worm, African ball worm.

1.2. Biology of *H. armigera*

The gram pod borer, *H. armigera* is polyphagous pest of gram, *Cicer arietinum* and red gram, *Cajanus cajana* and other plants.

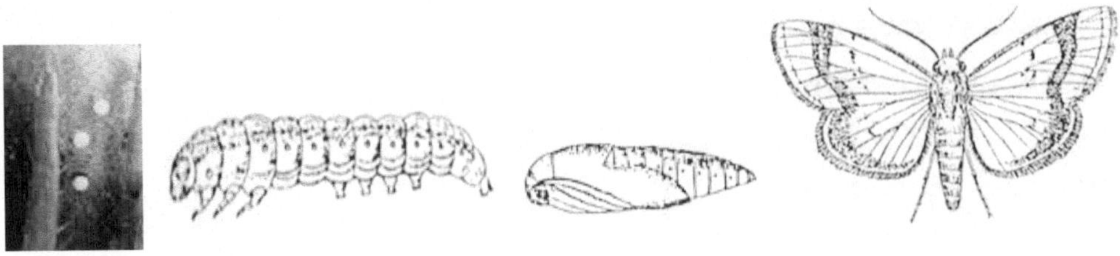

A B C D

Figure 8.1: *H. armigera*.
A: Eggs; B: Larva; C: Pupa; D: Moth.

Egg: The adult female moth deposits egg singly on tender leaves and flowers of plants. The female moth lays about 700 eggs. The freshly laid eggs are spherical, white and changes to dark brown before hatching (0.4 mm to 0.6 mm in width). The incubation period of eggs is about 4 days.

Larva: The first and second larval instars are yellowish-white to reddish-brown with dark brown to black head capsule. Three pairs of thoracic legs and five pairs of prolegs were developed in third instar stage larvae. The full grown larva is straw-yellow to green with lateral brown strips. The head and prothoracic legs are dark brown to black in color. Tubercles and spiracles of the larvae are also brown to black. The full grown larvae become sluggish, wrinkled with stopped feeding and movement. Newly hatched caterpillar first feeds on leaves for short time then it attacks the pods. The larva is greenish to brown, but variable in coloration. Larva has short hairs and dark stripes on the body. Body bears dark and pale bands. There are six larval instars. Larval developmental period is about 24 days.

Pupa: Pupation takes place in the soil. The brown pupa is about 16 mm long. The surface is smooth ends with two dark brown spines. The pupal stages last for about 10 days. The pest hibernates as pupae during cold months of January and February.

Moth: The adult moth is medium sized, stout-bodied, nocturnal, light brown and measures about 40 mm across the wings span. The moth has v-shaped speck (mark) on the light brownish forewings and hind wings are lighter in color with a dark patch at the outer end. Male is smaller than female. Tufts of hairs are present on the tip of the broad abdomen in female. Total life cycle from egg to adult takes about 37 days. The adult longevity is about 6 days and feed on plant nectar. There are about eight generations of the pest in a year. Cannibalism is found in the larvae of *H. armigera.* Biology studied at 28 °C ± 1 °C and 75-80 per cent humidity.

1.3. Damage Caused by *H. armigera*

A pest is an animal whose population often increase above a certain level of economic injury and its existence conflict with man's welfare, convenience and profits. The insects which cause damage to crop plants are called agriculture pests. The pest is said to be economic pest if it causes 5 per cent or more loss to the crop.

The amount of damage caused to a crop is called economic damage (Edwards and Heath, 1964). There are three types of insects such as beneficial, harmful and neutral insects. The caterpillars cause damage to gram and red gram. This pest feeds on young pods, leaves (foliage) and seeds resulting in reduced yield of gram and red gram and other plants. It directly damages maize ear, cotton balls and tomato fruits. It also gives damage to the sunflowers. It is completely polyphagous (more than 181 plants) pest. It is also a minor pest on many cultivated fruits. In America it causes maximum destruction to cotton, hence there it is called as American cotton boll worm. The pest is cosmopolitan in distribution. It damages chickpea and yield loss was found to be 10 to 60 per cent. The third and fourth instars larvae cause maximum damage. The larva causes defoliage of gram plants and other plants. It also damages to cauliflower, sunflower, sweet potato, mango, tomato and pea plant (*Pisum sativum*). This pest is listed by EPPO (European and Mediterranean Plant Protection Organization).

1.4. Control of *H. armigera*

Insect pest can be control by main three methods such as mechanical, chemical and biological methods.

a. Collection and destruction of the larvae.

b. Installation of female moth pheromone traps will attract and kill male moths.

c. Installation of light traps attract for both sexes and moths killed in kerosene.

d. Sunflower + pigeon pea inter cropping is suggested for less incidence of *H. armigera.* Spary the chiili garlic extract or 5 per cent Neem seed kernal extract to control army worm.

e. Spraying methyl parathion 0.025 per cent or cypermethrin 0.025 per cent or fenevalerate 0.02 per cent or monocrotophos 0.04 per cent or quinalphos 0.025 per cent econeem or neem oil neem seed kernel extract 5 per cent), Bt (*Bacillus thuringenesis*) liquid formulation (2 liters/hector) are also effective for *H. armigera*. Spary the chiili garlic extract or econeem or neem oil (neem seed kernel extract 5 per cent), neam leaf extract (0.5 per cent) is suggested as preventive measure (Gillot, 1982, Nayar *et al.*, 1998, Antram, 1986, Atwal and Dhaliwal, 2003, Ravindranathan, 2003 and David and Ananthakrishnan, 2006).

f. The larvae have predators like preying mantis, *Crebrator gemmatus* and a yellow wasp, *Polistes herebreus.*

g. The larvae have predators like larvae and ladybird beetle adult of *Minochiles sexmaculatus.* h. The larvae have parasitoids like insect parasitoid wasp, *Apanteles ruficrus.*

i. The larvae have virus parasitoid like *Trichogramma chilonis.* For this Vasant Trichocards may be used.

g. HNPV (*Heliothis* nuclear polyhedral virus) are used for eggs and larval control.
h. Muscardiae disease develops in larvae and eggs of pest by using fungus.

2. *Papilio demoleus* (Linnaeus)

2.1. Systematic Position of *Papilio demoleus* (Linnaeus)

Kingdom : Animalia

Phylum : Arthropoda

Class : Insecta

Subclass : Pterygota

Division : Endopeterygota

Order : Lepidoptera

Family : Papilionidae

Genus : *Papilio*

Species : *demoleus* (Linnaeus)

Common name: Lemon butterfly; Citrus butterfly, Swallowtail butterfly.

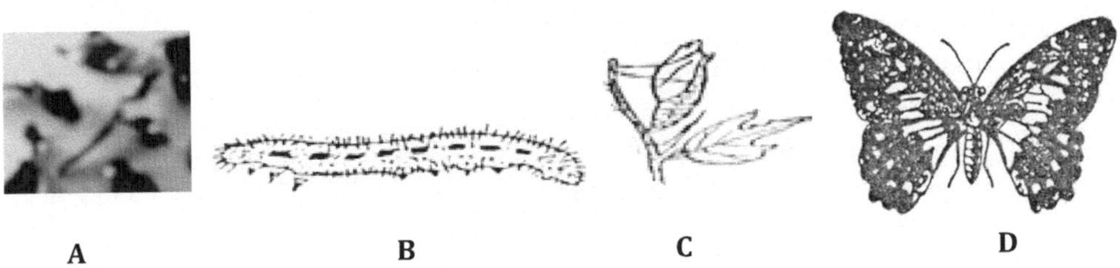

Figure 8.2: *P. demoleus*.
A: Eggs; B: Larva; C: Pupa; D: Butterfly.

2.2. Biology of *P. demoleus*

The lemon butterfly, *P. demoleus* is a pest of citrus.

Egg: A female lays about 120 grayish-yellow eggs on the underside of tender leaves. Egg incubation period is about 4 days during summer and 5 days during winter. The fertile egg (1.5 mm in length) develops a small red mark at the apex.

Larva: Larva bears thoracic legs, ventral prolegs, anal proleg and spiracles. There are five larval instars. Early instar larva is dark with white patches resembling the dropping of birds. When grown up it turns deep green in colour and measures 4 cm in length. It has hump-like raised portion anteriorly. The young caterpillar is brown with white markings. The colour changes to green with brown bands across the back when full grown. The larva has osmeterium which is yellow at the base and orange at the tip. Osmeterium is fleshy, glistening, bifid (forked) organ occurs just behind the head on the larvae. Osmeterium is normally hidden but can be everted when the larva feels threatened and emits pungent compounds that deter some predators. Larval developmental period is about 14 days.

Pupa: Pupation takes on the stem. The pupa is dimorphic with green and brown colours. It hangs on the stem by two fine silken threads secreted by it. Pupal period is about 11-days.

Butterfly: It is a beautiful butterfly of large size measuring about 2.5 cm in length and 9 cm on wing span (expanse). The forewing base has yellow scales, two large yellow spots and a curved spot at its apex. Hind wing base has yellow scales and yellow median bands and a series of smaller yellow spots at its apex. A large eye-like spot with blue scales. The upper hind wings of male and female butterfly have a red tornal spot. Black coloured antennae are clubbed. The dorsal side of head, thorax and abdomen is black and covered with yellow scales and ventral surface is cream coloured. This pest is active throughout the year but it hibernates in the pupal stage during severe winter. The longevity of the male adult is 4-days and female adult is 7-days. Adult male and female butterflies feed on plant nectar in the field. The butterflies usually appear in the morning and begin to fly within one to three hours of emergence. Adult emerged from pupa with the release of moulting fluid. The male is smaller than female butterfly. Four generations have been recorded in one year. Biology studied at 28 °C ± 1 °C and 75-80 per cent humidity.

2.3. Damage Caused by *P. demoleus*

The lemon butterfly, *P. demoleus* is very destructive pest of citrus. Larvae cause damage to citrus plants. The young caterpillar initially feeds on small leaves attacking larger ones as it grows older. Young larvae feed on fresh leaves and terminal shoots. Larvae feed voracious for about 8-16 days in summer. The caterpillars cause defoliage and destruction of the citrus plants. It has been observed that they start feeding from margin and reach the midrib. The severe attack by full grown caterpillars causes complete defoliation of citrus plants thus rendering them unfit for fruit bearing. Apart from *citrus* sp. it also attacks on Kadi patta plant, *Murraya koenigii;* Bel plant, *Aegle marmelos;* Kavatplant or wood apple or Elephant apple, *Ferontia elephantum* and other rutaceous plants.

2.4. Control of *P. demoleus*

Insect pest can be control by main three methods such as mechanical, chemical and biological methods.

a. Hand picking and destruction of the larvae in nurseries and field.
b. Installation of female moth pheromone traps will attract and kill male moths.
c. Installation of light traps attract both sexes and moths killed in kerosene or oil.

d. Spraying methyl parathion 0.025 per cent or cypermethrin 0.025 per cent or fenevalerate 0.02 per cent or monocrotophos 0.04 per cent or quinalphos 0.025 per cent econeem or neem oil neem seed kernel extract 5 per cent), Bt (*Bacillus thuringenesis*) liquid formulation (2 liters/ hector) are effective for *P. demoleus*. Spary the chili garlic extract or 5 per cent Neem seed kernal extract is suggested as preventive measure (Gillot, 1982, Nayar *et al.*, 1998, Antram, 1986, Atwal and Dhaliwal, 2003, Ravindranathan, 2003 and David and Ananthakrishnan, 2006).

e. The larvae are attacked by preying mantis, *Crebrator gemmatus* and a yellow wasp, *Polistes herebreus,* wasp, *Pteromalus luzonensis* (Pteromalidae) and wasp, *Telenomus remus* (Scelionidae is Subfamily of Pteromalidae).

f. The larvae have parasitoids like *Apenteles* (*Ooencyrtus*) *papilionis, Apenteles* sp., *Bracon hebetor, Charops* sp. (Ichneumonidae).

g. The pupae have parasitoids like wasps, *Brachymeria nephantidis* (Chalcididae h. The egg parasitoids associated with this pest are *Trichogramma evanescens* Westwood (Trichogrammatidae), *Pteromalus luzonensis* (Pteromalidae) and *Telenomus* sp. (Scelionidae).

3. *Pyrilla perpusilla*

3.1. Systematic Position of *Pyrilla perpusilla*

Kingdom : Animalia

Phylum : Arthropoda

Class : Insecta

Subclass : Pterygota

Division : Endopeterygota

Order : Hemiptera

Family : Lophopidae

Genus : *Pyrilla*

Species : *perpusilla*

Common name: Sugarcane leafhopper; Sugarcane plant bug:

3.2. Biology of *P. perpusilla*

Sugarcane planthopper, *P. perpusilla* is a major pest of sugarcane.

Egg: Females lay eggs on the lower side of the leaves near the midrib. Eggs are laid in clusters of 30-40 in number in rows of 4-5. They are covered by pale waxy material. Eggs are oval shaped, pale whitish to bluish green when laid and turn brown before hatching. The females bear a pair of anal tufts which produce waxy filaments to cover the eggs. A female lays about 500 eggs. Egg incubation period is about 8 days.

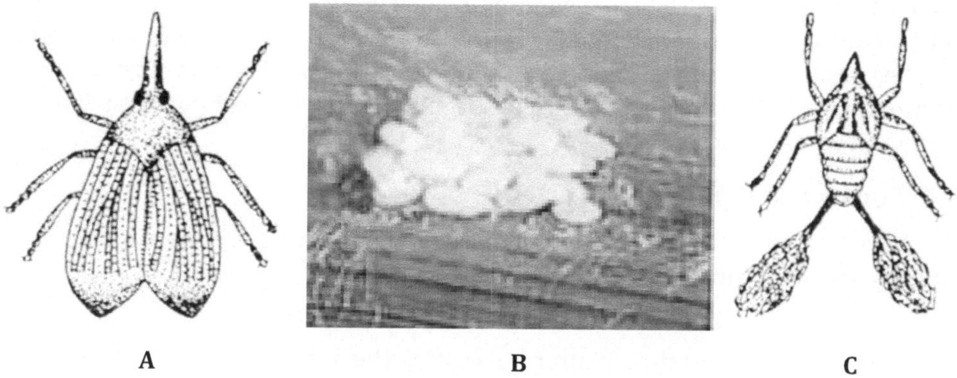

Figure 8.3: ***P. perpusilla*.**
A: Bug; B: Eggs; C: Nymph.

Nymph: Nymph passes through five instar stages to reach adult stage. First instar nymph has creamy-white colour. Nymph has white colored thread on last abdominal segment. Second instar nymph has dark brown strips along the lateral margin. Third instar nymph has a thoracic region with one dark brown dorsal patch on either side. Abdominal segments are pale blue on dorsal side and pale yellow on ventral side. Abdominal segments of fourth instar nymph have a dark brown band on the dorsal surface and pale green ventral surface. Abdominal segments of fifth instar nymph have a dark brown transverse band on the dorsal surface and pale white ventral surface. Nymphal period is about 32 days.

Bug: Bugs are yellowish-brown and later turns straw colored. Eyes turn pale green and head develops a snout with a black spot at the posterior side. Piercing and sucking mouth parts present in adult. Male measures about 20 mm in wingspan. Male is smaller than female. Adult longevity is about 52 days. Female lays eggs mainly during the day. Head shows narrow portion called as a snout. The wingspan of adult bug is 2 cm. Adult of *Pyrilla* is called bug. Fore wings are semi-opaque in nature with yellow-brown color. The fore wings are lightly covered with minute black spots. On the slightest disturbance, they jump from leaf to leaf. Adult *P. perpusilla* is a pale yellow and soft body insect. It is cosmopolitan.

3.3. Damage Caused by *Pyrilla perpusilla*

The damage is done by nymphs and adult bugs. Adults and nymphs suck sap from the under surface of leaves of sugarcane and other plants. *Pyrilla* causes great loss to the yield of sugar cane. Honeydew secreted by nymphs cause sooty mould and affect photosynthetic activity. Leaves wither and fall off. Appearance of streaks due to punctures made by bugs and nymphs. Sugar cane yield get reduced upto 50 per cent. *Pyrilla* heavy infested leaves turn yellowish white. It causes difficulties in milling of cane. Sucking of sap from the leaves weakens the plant and reduces the sucrose contents by up to 50 per cent. The hoppers secrete a sweet sticky fluid called "honeydew" which promotes a quick growth of fungus *Capnodium* sps. Sooty mould grows on honeydew. It results into decrease in photosynthesis. Brown patches occurred on the leaves. Leaves may desiccated and become brown. It also gives damage to the maize, bajra, rice, sorghum, okra, watermelon, pea, bamboo plant and wild grasses.

3.4. Control of *P. perpusilla*

Insect pest can be control by main three methods such as mechanical, chemical and biological methods.

a. Collection and destruction of the nymph and adults.
b. Burning of all thrash after harvesting the sugarcane crop in an infected field.
c. Removal of leaves bearing egg clusters in October planted crop in March-April.
d. Installation of female bug pheromone traps will attract and kill male bugs.
e. Installation of light traps attracts male and female bugs and killed in oil.
f. The burning of trash helps in destroying nymphs and unhatched eggs.
g. Dusting the infested crop with 10 per cent BHC, 10 per cent Toxaphane and 10 per cent Carbaryl or 5 per cent Melathione @ 20-40 kg/ha.
h. Sprayings of the insecticides like parathion (2 per cent) at 12.5 kg/ha, fenthion (560 mL/ha); malathion (500 mL/ha); phosphomidon (250- 300 mL/ha); monocrotophos (1250 mL/ha) and Bt (*Bacillus thuringenesis*) liquid formulation (2 liters/hector) and econeem or neem oil (neem seed kernel extract 5 per cent) is also effective for this pest.
i. The light and pheromones traps are used for the control of this pest.
j. The ectoparacitoid insect moth, *Epiricania melanoleuca* is biological control measure for this pest. The cocoons of moth, *Epiricania melanoleuca* is released on the leaves for biological control this pest.
k. The eggs have bacterial parasitoids like *Trichogramma evanescens*.
l. The eggs virus parasitoids like *Trichogramma chilonis* (50, 000/ha) are effective for this pest.
m. The white muscardine fungus is effective for this pest.

4. *Tribolium castaneum*

4.1. Systematic Position of *Tribolium castaneum*

Kingdom : Animalia

Phylum : Arthropoda

Class : Insecta

Subclass : Pterygota

Division : Endopeterygota

Order : Coleoptera

Family : Tenebrionidae

Genus : *Tribolium*

Species : *castaneum* (Herbst, 1797)

Common name: Red flour beetle, Red rust flour beetle, Meal beetle, Darking beetle.

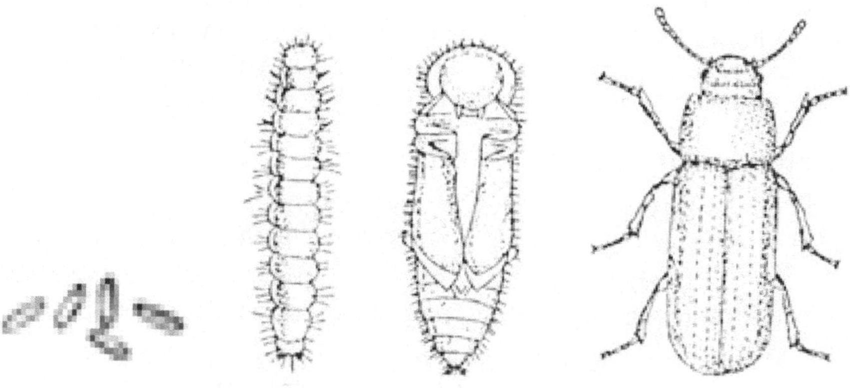

A B C D

Figure 8.4: *T. castaneum*.
A: Eggs; B: Grub; C: Pupa; D: Beetle.

4.2. Biology of *T. castaneum*

T. castaneum is a pest flour of wheat and also a pest of stored grains.

Egg: Eggs are white, microscopic and lay amongst the grain or in the flour. Female beetle lays about 24 eggs. Eggs are cylindrical and sticky, therefore flour easily get adhere to the egg surface. Eggs (0.5 mm in diameter) have bits of flour stuck to their surface. Egg incubation period is about 4-5 days.

Grub: The slender grubs are creamy yellow to light brown in colour. Grubs have dark brown eyes and body covered with fine hairs. They have two dark, upturned, pointed projections on the last body segment. Grubs remain hidden in the flour. Grub head distinct, black and sclerotized. Prolegs short and inactive. Grub is about 0.6 cm in long. Grub developmental period is about 28 days with seven instars.

Pupa: Pupation occurs in flour. The pupae are yellowish and hairy. They can be found on wheat flour, wheat, cereal products, pulses, dry fruits *etc.* Pupa (0.4 cm) is white to yellowish in colour with dark wings, sclerotized legs and fully developed eyes. Pupa did not have cocoon and initially it is white then turn to yellow and then becomes brown. Pupa covered with fine hairs. It is non feeding stage. Pupal period is about 5-9 days. Male (0.3 cm) is smaller than female (0.37cm).

Beetle: Beetles are reddish brown in colour. Each antenna enlarges gradually towards the tips. Antennal last three segments (club with 3 segments) are gradually increased. Beetle has chewing mouth parts. These are small sized 0.5 cm in length. Thorax is slightly curved lateral sides. Beetle may cause allergic reactions. Last abdominal segments dark. They do not damage the furniture. They do not bite or sting. Beetle may fly and attract towards light. In other species of *Tribolium confusum* last four segments (club with 4 segments) are equal in size, thorax is straight. Head dark, notched eye present. Wings are non-functional. Beetles are found in warm places. Male has small patch on the fore femur. Adult male beetle longevity is about 12 months and female beetle longevity is more than 15 months. Female is polyandrous (mating with more beetles). Adults are long lived and may live upto three years. Life cycle requires 40-90 days. They are cosmopolitan in distribution.

4.3. Damage Caused by *T. castaneum*

a. Adults are incapable of feeding on sound grains but they do considerable damage to flour and flour products and to grains damaged by other pests.
b. Beetle attacks flour of wheat, stored grains and other food products including cereals, pasta, biscuits, beans, dried fruits and nuts.
c. Damage is done by beetles and grubs during humid seasons.
d. Powdery grains and pungent odour (foul smell) make flour/grains unfit for human consumption.
e In severe infestation flour turns greyish and mouldy. Flour becomes unfit for human consumption. It causes a grey tint to the grain.

4.4. Control of *T. castaneum*

Insect pest can be control by main three methods such as mechanical, chemical and biological methods.

1. Collection and destruction of the grubs and beetles.
2. Female beetle pheromone traps that will attract and kill male adult beetles.
3. Light traps for both sexes attract beetles and killed in kerosene or oil.
4. Clean and dry the store house.
5. Deep freezing of flour and grains.
6. Use resistant varieties of crops.
7. Use Gamma radiation dose to sterilize beetle.
8. Mix bioinsecticide roots of kavali plant, *Decalepis hamiltonii*.
9. Use malathion 5 per cent at the rate 250 g/quintal seeds.

10. Mixing of fevalerate or 1.5 mL of cypermethrin or 14 mL of deltamethrin per quintal of spreading seeds by diluting in 500 mL of water.
11. Mix neem leaves in the stored graina.
12. Fumigation with aluminium phosphide protects the seed.
13. Exposure of the grain to sun light.
14. Repackaging material into new infection-free containers.
15. Placing the products in the freezer for several weeks may help to kill weevils and grubs.

5. *Callasobruchus chinensis*

5.1. Systematic Position of *Callasobruchus chinensis*

Kingdom : Animalia

Phylum : Arthropoda

Class : Insecta

Subclass : Pterygota

Division : Endopeterygota

Order : Coleoptera

Family : Chrysomelidae or Bruchidae

Genus : *Callasobruchus* (*Pachymerus*)

Species : *chinensis* (Linnaeus, 1758)

Common name: Pulse beetle, gram dhora beetle, Chinese bruchid beetle, Cowpea bruchid beetle, Adzuki bean weevil.

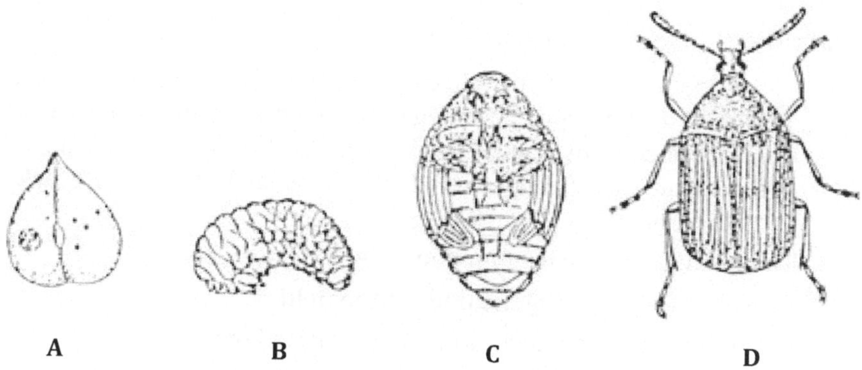

Figure 8.5: *C. chinensis*.
A; Eggs; B: Grub; C: Pupa; D: Beetle.

5.2. Biology of *C. chinensis*

Pulse beetle, *C. chinensis* is a pest of pulses gram, mung (*Phaseolus aureus*), peas, soyabean and cowpeas. It is also the pest of cotton, sorghum and maize seeds.

Egg: Eggs are scale like and glued to the grains. Eggs white, microscopic and lay on the grains. Female beetle lays about 100 eggs. Eggs are sticky, therefore grains easily get adhere to the egg surface. Eggs (0.5 mm in diameter) have bits of grains stuck to their surface. Eggs hatch in 3-5 days.

Grub: The slender grubs are creamy yellow to light brown in colour. Grubs have dark brown eyes, mouth parts black and body covered with fine hairs. They have two dark, upturned, pointed projections on the last body segment. Grubs are white, creamy, curved, remain hidden in the flour. Grub (full grown 0.6 cm) period is about 14-28 days with seven instars.

Pupa: Pupae are oval and white. They can be found on pulses and grains. Pupa (0.4 cm) is white to yellowish in color with dark wings, sclerotized legs and fully developed eyes. Pupa did not have cocoon but lies inside the surrounded grains and seed coat. Pupa covered with pulses and grains. Pupa is non feeding stage. Pupal period 4-28 days. Male (0.4 cm) is smaller than female (0.5 cm).

Beetle: Beetles are small, oval, reddish brown colored with blunt snout. Antenna is long, serrated and gradually enlarges towards the tips. Beetle has chewing mouth parts. Beetles are oval, small sized 0.4 cm in length. There is a pair of white elongate prominences in the middle of the hind margin of the thorax. Beetle may cause allergic reactions. They do not damage the furniture. They do not bite or sting. Beetle may fly and attract towards light. Wings (elytra) not reach up to the tip of abdomen. Beetles are found in warm places. Hind femur with spine at inner and outer sides. Hind femur of female is thickened. Male and female beetle longevity is about 7-8 days. Life cycle requires 40-90 days.

5.3. Damage Caused by *C. chinensis*

1. Pulse beetle is a pest of pulses gram, mung (*Phaseolus aureus*), peas, soyabean and cowpeas. Damage is caused by grubs and they eat the entire seed content leaving outer covering the seed only. In the field the eggs are laid on green pods. Pulses and seeds become unfit for human consumption.
2. The maximum damage found in April to September.
3. Beetles start attacking the pods in the field.
4. From the field, the infestation will be carried over to the godowns.
5. If the godown is poorly aerated, fungus and foul smell could also develop
6. Presence of holes on the grains. Flour of infested pulse has a characteristic off- flavour and should not be marketed.

5.4. Control of *C. chinensis*

Insect pest can be control by main three methods such as mechanical, chemical and biological methods.

1. Hand picking and destruction of the larvae may be done.
2. Clean and dry the store house.
3. Deep freezing of flour and grains.
4. Use resistant varieties of crops.
5. Use Gamma radiation dose to sterilize beetle.
6. Use malathion 5 per cent at the rate 250 g/quintal seeds.
7. Mixing of fenvalerate (it is synthetic pyrethroid) or 1.5 mL of cypermethrin or 14 mL of deltamethrin per quintal of spreading seeds by in 500 mL water.
8. Mix neem leaves in the stored graina.
9. Fumigation with aluminium phosphide protects the seed.
10. Exposure of the grain to sun light.
11. Repackaging material into new infection-free containers.
12. Placing the products in the freezer for several weeks may help to kill weevils and grubs.

6. *Sitophilus oryzae*

6.1. Systematic Position of *Sitophilus oryzae*

Kingdom : Animalia

Phylum : Arthropoda

Class : Insecta

Subclass : Pterygota

Division : Endopeterygota

Order : Coleoptera

Family : Curculionidae

Genus : *Sitophilus*

Species : *oryzae* (L.).

Common name: Rice weevil.

6.2. Biology of *Sitophilus oryzae*

Rice weevil, *S. oryzae* is a pest of stored rice and other store grains.

Egg: Rice weevil lays its eggs in crevices of kernels or dust. Female rice weevil lays about 4 eggs/day. During its life span of 5 months, it lays a total of about 250-

400 eggs. The eggs hatch in 4-6 days.The eggs are laid in the cavities made on the grain by the mandibles of the weevils.

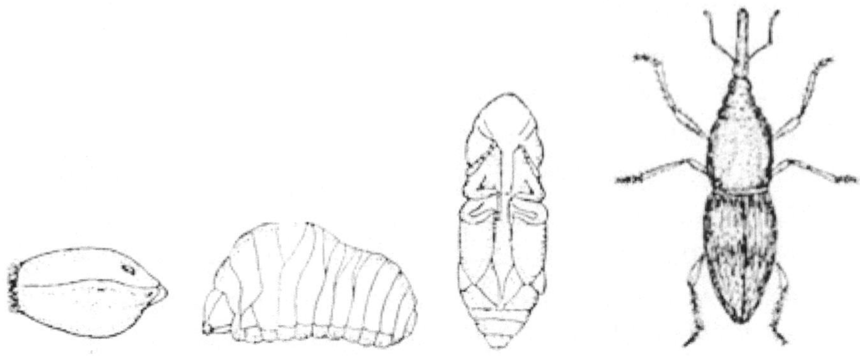

A B C D

Figure 8.6: *S. oryzae*.
A: Eggs; B: Grub; C: Pupa; D: Weevil.

Grubs: They feed inside the grain kernel for 18 days. Inside the seed, larval cuticle hardens and matures. The grubs (5mm) are white with yellowish brown head. Young grubs bore into the grains, eat their kernels, grow and pupate inside them. Grub is white and head yellow-brown in colour. Grub is legless 5 mm in length. Grub developmental period is about 25-30 days.

Pupa: The pupa stage lasts for 4-5 days. The pupa does not feed. In some species, the pupa is enclosed in a cocoon constructed by the larva. The pupa goes through great changes both internally and externally. Finally the insect emerges out as a developed adult. Pupa is movable, white and turns to brown.

Weevil: Weevils are between 3 mm to 4 mm long. They have three pairs of legs and their bodies are divided into head, thorax and abdomen. Weevils move and penetrate deeply into bulk of grains and get widely distributed. They are cosmopolitan in distribution. They are reddish brown to black, four yellow or reddish spots present in rounded depression on the wing (elytra). Weevils in winter season found in cracks and crevices of the godowns and store grain room. It has a long and curved rostrum. Adult is active flier. When disturbed, adult pull in their legs and fall to the ground and look like dead. The longevity of the adult weevil is 6 months. Adult lays about 400 eggs. Eggs get sealed by the female beetles on the seeds. Biology is studied at 28 °C ± 1 °C temperature and 75-80 per cent relative humidity.

6.3. Damage Caused by Rice Weevil, *S. oryzae*

Rice weevil, *Sitophilus oryzae* is a pest of stored rice. Adults and grubs also damage to the maize, wheat, corn, cotton seeds, barley and sorghum. It makes round holes to the grains and rice. Damage is noted in increased moisture. It prevents seed germination. Grubs develop pieces of rice seeds. Large infestation increases in the temperature of the stock by breeding in the excrement and may cause mold. The damage caused by them may extent up to 50 per cent of the total grain stored in a particular godown. Both adults and larvae feed upon the grains and make them inconsumable. They can actually cause large losses since the female makes holes somewhere in the plant, either branches or fruits to oviposit. It also gives damage to the wheat, cotton seeds, maize and sorghum.

6.4. Control of *S. oryzae*

1. Collection and destruction of the weevils. Removal of infested seeds.
2. Female weevil pheromone traps will attract and kill male adult weevils.
3. Light traps attract both sexes and weevils killed in kerosene or oil.
4. Use 0.05 per cent malathion (Organophoshphates). It is contact and stomach poison. It blocks nerve impulse and causes paralysis and death.
5. Use 0.15 per cent carbaryl. It is contact and systemic action.
6. Use aerosol like phantom for killing grubs.
7. Use pest proof container. Use vacuum pump and clean store room.
8. Heating with 60 °C for 15 minutes. Deep freezing below -7.5 °C is effective.
9. Use dry ice (which contains more CO_2 concentration) for preservation of grains and rice. Use insect ectoparasitoid like *Anisopteromalus calandrae* (H.) for control of weevils.
10. Mix neem leaves in the stored graina.
11. Fumigation with aluminium phosphide protects the seed.
12. Exposure of the grain to sun light.
13. Repackaging material into new infection-free containers.
14. Placing the products in the freezer for several weeks may help to kill weevils and grubs.
15. Freeze rice for at least one week to kill eggs.

Study of Insect Damage to different Plant Parts/Photographs

1. *Leucinodes orbonalis* (Guenee)

Class : Insecta

Subclass : Pterygota

Division : Endopeterygota

Order : Lepidoptera

Family : Pyraustidae

Genus : *Leuconodes*

Species : *orbonalis* (Guenee)

Common name: Brinjal shoot and fruit-borer.

A B C

Figure 8.7: ***L. orbonalis.***
A: Larva; B: Pupa; C: Moth.

Moth lays about 250 eggs. The egg incubation period is about 3-4 days.The larvae grow through 5 stages and are full-fed in 11 days. The pupal stages last about 10 days. The adult longevity is about 5 days. Larvae bore into tender shoots near the growing points. Larvae bore into flower buds. Larvae bore into the fruits. When tender shoots are attacked, the growing points are killed. The infested fruits decrease market value. It is not unknown to see the whole of the crop destroyed by the bores. It also feeds on other solanaceous plants. It occasionally feeds on green pods of peas.

2. *Earias vittella*

Class : Insecta

Subclass : Pterygota

Division : Endopeterygota

Order : Lepidoptera.

Family : Arctiidae

Genus : *Earias*

Species : *vittella* (Fabricious).

Common name: Spotted bollworm.

It is the pest of okra. It is widely distributed in North Africa, Indo-Pakistan, sub-continent and other countries of world. Female moth of *E. vittella* deposits about 300 eggs on okra pods. The egg incubation period is about 4 days. Larva measures about 1.8 cm in long. Larval developmental period is about 17 days. The pupation period of *E. vittella* was noted as 11 days. The adult longevity is 16 and 18 days in male and female moths respectively. Larval stage is destructive and it

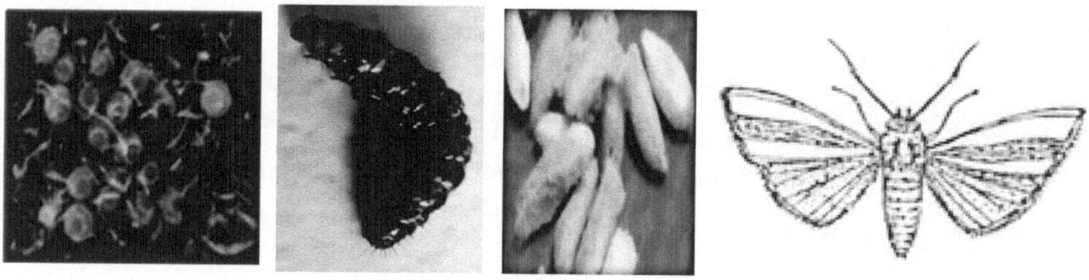

A B C D

Figure 8.8: ***E. vittella.***
A: Eggs; B: Larva; C: Pupae; D: Moth.

attacks growing points, flowers and pods. The freshly hatched larvae bore into tender terminal shoots, buds, flowers and pods. The caterpillars bore inside the pods and feed on inner tissue. They move from bud to bud and pod to pod thus causing damage to the number of fruiting bodies. The damaged buds and flowers wither and fall down without bearing any pods, whereas the affected fruits become deformed in shape and remain stunted in growth and shows drying. They are active almost throughout the year on different host plants under field condition.The larvae of *E. vittella* cause damage to more than 180 crops resulting in 69 per cent reduction in yield. The infested fruits become unsuitable for human consumption. The bored holes present on the pods are seen plugged and excreta thus quality of the fruit reduced (Sewak, 2016).

3. *Helicoverpa armigera.*

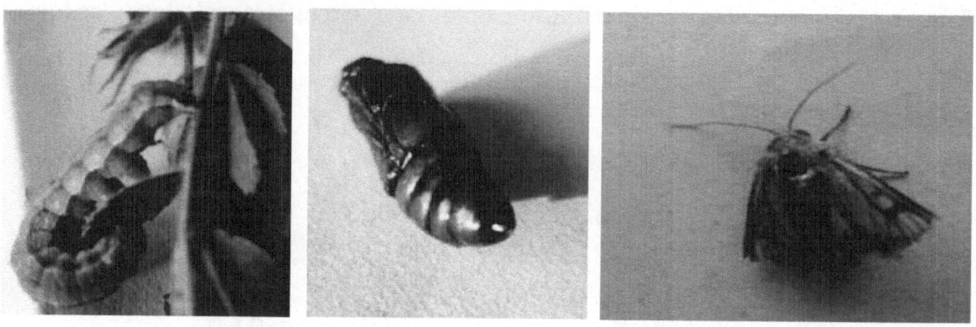

A B C

Figure 8.9: ***H. armigera***
A: Larva; B: Pupa; C: Moth.

The gram pod borer, *H. armigera* is a pest of gram plant. There are three types of insects such as beneficial, harmful and neutral insects. The caterpillars feed on young pods, leaves (foliage) and seeds. It results into the reduction in yield of gram. Newly hatched caterpillar first feeds on leaves for short time then it attacks the pods. It directly damages to the leaves and pods of gram. It is polyphagous pest (more than 181 plants). The pest is cosmopolitan in distribution. It damages chickpea and yield loss was found to be 10 to 20 per cent. The third and fourth instars larvae cause maximum damage. The larva causes defoliage of gram plants.

4. *Papilio demoleus*

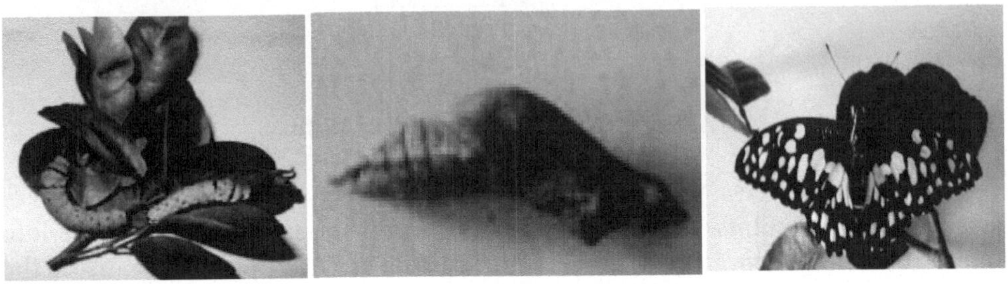

A B C

Figure 8.10: *P. demoleus*.
A: Larva; B: Pupa; C: Butterfly.

The lemon butterfly, *P. demoleus* is a pest of citrus plant. It causes damage to citrus plants. The young caterpillar initially feeds on small leaves attacking larger ones as it grows older. It feeds for about 8-16 days in summer. The caterpillars cause destruction of the citrus plants. Young larvae feed on fresh leaves and terminal shoots. It has been observed that they start feeding from margin and reach the midrib. It causes defoliation of citrus plants. Plant becomes unfit for fruit bearing. The pest is cosmopolitan in distribution. It damages citrus and yield loss was found to be 20 to 30 per cent.

5. *Pyrilla perpusilla*

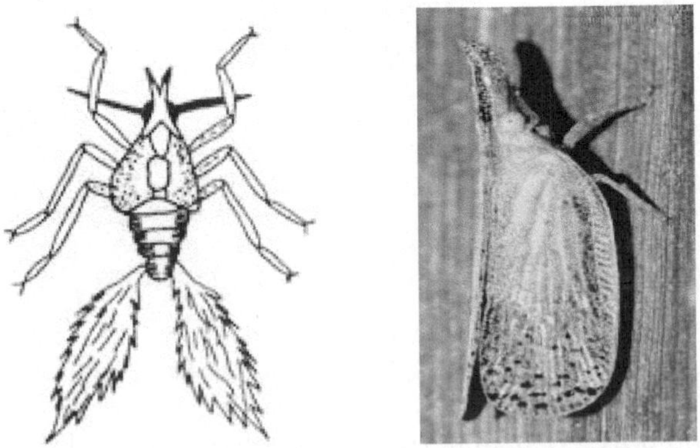

A B

Figure 8.11: *P. perpusilla*.
A-B: Nymph; B: Bug.

Sugarcane planthopper, *P. perpusilla* is a pest of sugarcane. The damage is done by nymphs and adult bugs. Adults and nymphs suck sap from the under surface of leaves of sugarcane and other plants. *Pyrilla* causes great loss to the yield of sugar cane. Honeydew secreted by nymphs cause sooty mould and affect photosynthetic activity. Leaves wither and fall off. Appearance of streaks due to punctures made by bugs and nymphs. Sugar cane yield get reduced upto 30 to 40 per cent. *Pyrilla* heavy infested leaves turn yellowish white. It causes difficulties in milling of cane. Sucking of sap from the leaves weakens the plant and reduces the sucrose contents by up to 50 per cent. The hoppers secrete a sweet sticky fluid called "honeydew" which promotes a quick growth of fungus *Capnodium* sps. Sooty mould grows on honeydew. It results into decrease in photosynthesis. Brown patches occurred on the leaves. Leaves may desiccated and become brown. The pest is cosmopolitan in distribution. The pest is cosmopolitan in distribution. The yield loss was found to be 30 to 40 per cent.

Insect Damage to different Stored Grains

1. Red Flour Beetle, *Tribolium castaneum.*

A B

Figure 8.12: *T. castaneum*.
A: Grub; B: Beetle.

Rust red flour beetle, *T. castaneum* is a pest flour of wheat. Adults are incapable of feeding on sound grains but they do considerable damage to flour and flour products and also to grains damaged by other pests. Beetle attacks flour of wheat. Damage is done by beetles and grubs during humid seasons. Pungent odour (foul smell) makes flour unfit for human consumption. In severe infestation flour turns greyish and mouldy. Flour becomes unfit for human consumption. It causes a grey tint to the grain. The pest is cosmopolitan in distribution. The yield loss was found to be 40-50 per cent.

2. Pulse Beetle, *Callasobruchus chinensis*

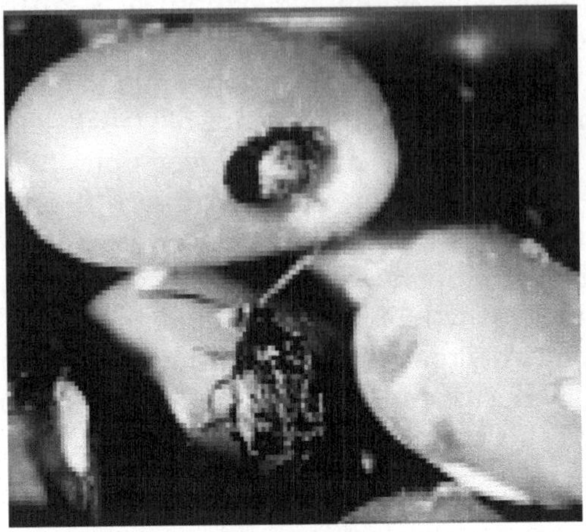

Figure 8.13: *C. chinensis.* Beetle.

Pulse beetle, *C. chinensis* is a pest of pulse gram. Damage is caused by grubs and beetles. They eat the entire seed content leaving outer covering the seed only. In the field the eggs are laid on green pods. Pulses and seeds become unfit for human consumption. The maximum damage found in April to September. Beetles start attacking the pods in the field. From the field, the infestation will be carried over to the godowns. It creates holes on the grains. Infested grains have a characteristic off-flavour and should not be marketed. If the godown is poorly aerated, fungus and foul smell could also develop. It causes a grey tint to the grain. It is cosmopolitan in distribution. The yield loss was found to be 50-60 per cent.

3. Rice Weevil, *Sitophilus oryzae*

Figure 8.14: *S. oryzae.* Weevils.

Rice weevil, *S. oryzae* is a pest of stored rice grains. Adults and grubs make rounded holes to the grains of rice. Damage is more in increased moisture. Infested grains have a characteristic off-flavour and should not be marketed. If the godown is poorly aerated, fungus and foul smell could also develop. The pest is cosmopolitan in distribution. It also gives damage to the wheat, cotton seeds, maize and sorghum. It causes a grey tint to the grain. The yield loss was found to be 60 to70 per cent. They eat the entire seed content.

Chapter

9

Insect Vectors of Human Diseases

9.1. Head Louse or Body Louse, Human Louse, *Pediculus humanus* (Linnaeus)

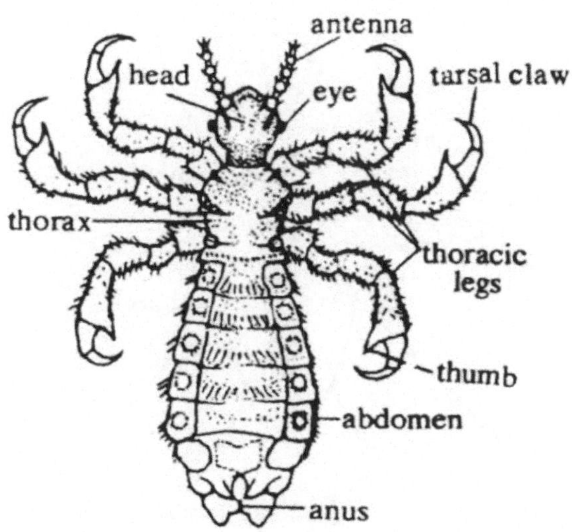

Figure 9.1: *P. humanus.* Adult.

Life cycle: Head louse, *P. humanus* belongs to class insecta and order anopleura. Wings absent. It is ectoparasite on human. It measures about 1-2/mm in length. It is wingless, dorsoventrally flattened and elongated insect with piercing and sucking mouth parts. They have three pairs of legs. Mandibles are rudimentary. Female head louse lays about 50 eggs (nits). Antenna five segmented. It crawls about 23/cm/minute. The egg cases (nits) are cemented to the base of hair. Piercing and sucking mouth parts present. Throracic segments fused and abdomen 9 segmented. Legs with single curved claws and opposite thumb for holding the hairs of host. Famale with narrow abdomen and does not turn upward. Egg incubation period is about 7 days. Louse has longevity of about 16 days. Larva is called nymph.

Vector of Disease: The body louse is the vector of three human diseases such as typhus, caused by *Rickettsia prowazeki;* trench fever in Africa caused by *Bartonella* (*Rochalimaea*) *quintana* and relapsing fever caused by *Borrellia recurrentis.* It feeds on blood of human. Body lice are a nuisance in themselves and cause intense itching. They spread epidemic typhus, trench fever and louse-borne relapsing fever. Body lice infestation is called pediculosis.

Control: Improve the personal hygiene of the infested person. Change clean clothes, bedding and towels. Cloth should be laundered using hot water (at least 130 °F or 54 °C) and machine dried using the hot cycle. Treat with a pediculicide (a medicine that can kill lice). Dust with 1 per cent malathion or 1 per cent permethrin is also effective.

9.2. House Mosquito, *Culex pipiens*

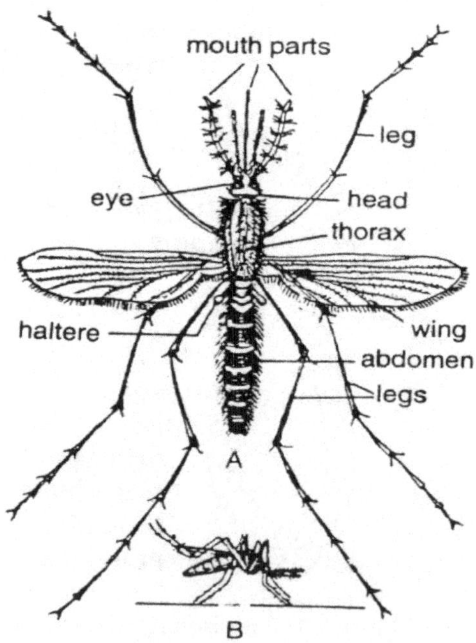

Figure 9.2: ***C. pipiens.***
A: Dorsal view; B: Sitting position.

Life Cycle: *Culex* mosquito belongs to class insect and order diptera. A pair of wings and a pair of halters present. It has a blunt tip on its abdomen. The male

Culex (feathery antenna) mosquito feeds on plant nectar. The body is parallel to the surface at rest. Wings with uniform color and unspotted. Body soft and covered with small scales. Legs without hairs. Proboscis is not in a line with body. Proboscis is shorter in male than female. Maxillary palps as long as proboscis in males; shorter three jointed in females. Maxillary palp is shorter in female than male. Antenna is stout in male and slender in female. Maxillae shorter than proboscis in female. Culicine female requires a blood meal. The eggs are laid on the surface of standing water present cans, garden pots, used tires and barrels. The female deposits about 300 eggs at a time. Egg incubation period is about 2 days. The larva will live in the water for about 14 days. Pupa floats on the water for about 4 days and transforms into an adult *Culex* mosquito. The female mosquito after mating feed on blood and lays its eggs. Proboscis and maxillae equal in male. Proboscis ends with labium and labrum. Brownish yellow basal bands present in the abdomen. Proboscis not in a line with the body. Body grey-black. In female small maxillary palps. Body remains parallel to surface in adult. Larva is bottom feeder and hangs from the water surface.

Vector of disease: *Culex* mosquito is not a primary vector of mosquito borne diseases such as malaria, dengue and yellow fever but it transmits a number of other illnesses that can present serious health problems to human beings. It is known to contribute to the spreading of West Nile Virus. Its bite is itchy and can be painful. *Culex* mosquito acts as vector for transmitting nematode pathogen, *Wruchereria bancrofti* which cause filariasis to humans. It is vector for transmitting a kind of virus for chikunguinea to man. It is vector for transmitting for encephalitis disease.

Control: A vaccine to provide protection from fever exists. Prevent mosquito bites include insecticides. Use mosquito traps, insect repellents and mosquito nets.

3. *Anopheles gambiae*

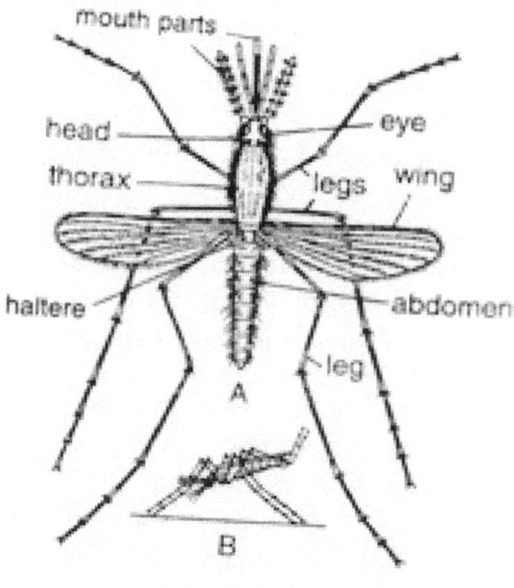

Figure 9.3: ***A. gambiae.***
A: Dorsal view; B: Sitting position.

Life Cycle: *Anopheles* mosquito (364 sp. in world and 51 sp. in India) belongs to class insect of order diptera. Body remains an angle with surface on which it rests. Proboscis is straight line with body. In female maxillary palps equal to labium or proboscis; in male maxillary palpi are clubbed. Abdomen generally without scales or with few scales. Wings spotted, palpae of male clubbed and in female it is straight. Anophiline adult female lays about 200 eggs on the water. It has slender body and long legs. Proboscis is same length in male than female. Maxillary palp club shaped and equal to proboscis in male. Maxillary palp simple and nearly equal to proboscis in female. Antenna is long and stout in male and slender and short in female (Ghose and Manna, 2016). Mosquitoes are zoophilic, anthropholic and exophilic. The head, thorax and abdomen are in the same straight line. Egg incubation period is about 3 days. The larva has mouth brushes used for feeding. Larva is surface feeder and found parallel to water. The larvae feed on algae, bacteria and other microorganisms. They go below the surface only when disturbed. Pupa is comma-shaped when viewed from the side. The pupal stage lasts about 3 days. Egg to adult duration is about 10-14 days. Female takes a blood meal for the production of eggs. Males live for about a week, feeding on nectar and other sources of sugar. Females will also feed on sugar sources for energy, but usually require a blood meal. Mosquito attracts towards human smells. Female mosquitoes are particularly drawn to foot odours).

Vector of Disease: *Anopheles* mosquito is vector of malaria and yellow fever. *Plasmodium vivax* is blood parasite transmitted by female *Anopheles*, mosquito. Mosquito bites causes transfer of sporozoite stage of *Plasmodium vivax* in blood. Sporozoite stage of *P. vivax* in human blood multiply to form merozoite stage. Merozoite stage feeds on content of R. B. cells and become gametocytes. Mosquito sucks gametocytes from human blood and develops in it's gut. In malaria symptoms like shivering fever, joint pain, vomiting, retinal damage, convulsion, sudden coldness and sweating are seen in the patient. Malarial pathogens are *P. vivax, P. falsiparum, P. ovale* and *P. maleri*

Control: A vaccine to provide protection from fever exists. Prevent mosquito bites include insecticides. Use mosquito traps, insect repellents and mosquito nets (Ghose and Manna, 2016, Sathe and Awate, 2009).

4. *Aedes aegypti*

Life Cycle: *A. aegypti* Mosquito belongs to class insecta of order diptera. They are visually distinctive because they have noticeable black and white markings on their bodies and legs. Proboscic and maxillary palps equal in length in male and not clubbed. Maxillary palps shorter than proboscis and three jointed in female. Abdomen and legs black with white strips. Proboscic is black. A lyre shaped (U-shaped) mark on the back of the thorax. Wings unspotted. The body is parallel to the surface at rest. Dark hairs on the antenna. Wings and legs with black and white bands, Proboscic black. Larva is bottom feeder and attached to the aquatic plant. The *Aedes* mosquitoes have 4 life stages. Adult females lay about 50 eggs. Egg incubation period is about 5 days. Larval period is 3 days and pupal stage lasts for

10 days. Adult longevity is 4 to 5 weeks. White bands are present on the abdomen. Mosquitoes can live and reproduce inside and outside the home.

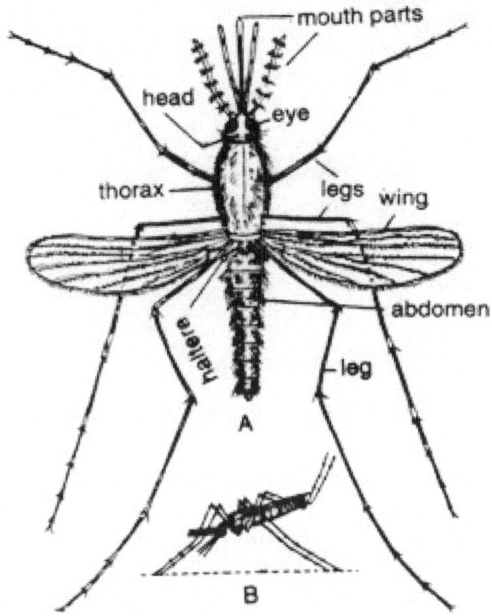

Figure 9.4: ***A. aegypti.***
A: Dorsal view; B: Sitting position.

Vector of disease: Unlike most other mosquitoes, they are active and bite only during the daytime. The peak biting periods are early in the morning and in the evening before dusk. Members of the genus *Aedes* are known vectors for numerous viral infections. The two most prominent species that transmit viruses are *A. aegypti* and *A. albopictus*, which transmit the viruses that cause, yellow fever, dengue fever West Nile fever, chikungunya, eastern equine encephalitis, and Zika virus. Infections with these viruses are typically accompanied by a fever, and in some cases, encephalitis, which can lead to death.

Control: A vaccine to provide protection from yellow fever exists. Use of mosquito bites like insecticides. Use mosquito traps, insect repellents and mosquito nets.

5. Rat Flea, *Xenopsylla cheopis* (Rothschild)

The rat flea, *X. cheopis* is an ectoparasite of rodents like genus *Rattus.* The rat flea (Pissu) belongs to class insect of order siphoneptera. It is small bilaterally compressed, wingless, short antena and hard skinned insect. Backwardly projected spines present. Large coxa present. Small head. Small thorax. Hind legs adapted for leaping. Abdomen 10 segmented. Sucking and piercing mouth parts present. Abdomen 10 segmented. Female flea lays about 400 eggs.

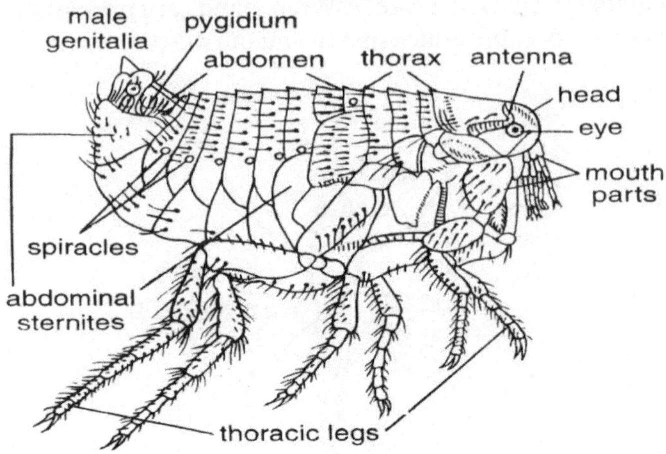

Figure 9.5: Rat Flea, *X. cheopis*. Adult.

Vector of disease: It is blood sucking ectoparasite. It causes itching and irritations. It is a parasite of rodents like genus *Rattus.* It is a primary vector of Bubonic plague, *Yersinia pestis, Rickettsia typhi* and murine endemic typhus.

Control: A vaccine to provide protection from fever exists. Use rat flea traps, insect repellents, nets and rat flea bites as in insects. Houses are treated with lindane 1 per cent or malathion 0.5 per cent or chlordane 2 per cent spray.

Chapter

10

Estrus Cycle of Rate

Estrus Cycle

Aim: Study of stages of estrus cycle of rat.

Principle

During breeding season the non-primate vertebrate female undergoes hormonal changes influencing reproductive organs, inducing histological changes leading to development of eggs. The cycle is called estrus cycle. The presence of nucleated epithelial cells, cornfield cells and leucocytes are the criteria for identification of phases of estrus cycle in rat.

Requirements

Compound microscope and mature female rat/permanent slides of estrus cycle.

Procedure

Take acclimatized mature female rat of 12-14 weeks age group. Give the treatment of ether or chloroform. Introduce through the vagina about 0.5 mL of normal saline solution (0.9 per cent) with the help of Pasteur pipette. Slowly draw out the vaginal swab with the same pipette. Put a drop of swab on the glass slide and make thin smear. Cover the semidried smear with methanol for fixing and leave for 10 minutes. Use nuclear stain like methylene blue or haematoxylin and cytoplasmic stain eosin. The presence of large nucleated epithelial cells, large cornfield cells and small rounded leucocytes are the criteria for identification of phases of estrus cycle in rat.

Phases of Estrus Cycle of Rat

Reproductive cycle of female rats is called estrous cycle. The beginning from first estrus phase to the nest estrus phase is called as estrus cycle. Female rat occur maturity after 30 to 50 days of age. The first estrus cycle begins one week after vaginal opening and thereafter regularly every about 5 days (120 hours). The estrus cycle stops during preganancy and does occurwithin 24 hours after birth. Animal having only one estrus cycle are called monoestrus. Those with many cycles in the same breeding season at short intervals are called polyestrus. The estrus cycle includes four phases such as proestrus, estrus, metestrus and diestrus. The duration of each phase depends on light, temperature and food. Rat is a polyestrus animal. In many species there are estrus cycle of varying duration, *e.g.*, 4–5 days in rat and mouse, 14 days in guinea-pigs, 16 days in sheep, 20 days in cows and twice a year in bitches, during which there is a special period of desire in the female accompanying the process of ovulation. Only at this time mating is possible (Keele and Neil, 1964, Nagbhushanam *et al.*, 1996, Ghose and Manna, 2016 and Srinivasan *et al.*, 2017).

Observation

1. *Proestrus Phase*

It is also called preparatory phase. It is predominant with large numerous nucleated epithelial cells. The development of ovarian mature follicles occur. It lasts for about 12 hours. Fluid collects in the uteri and they become highly contractile. Copulation is not permitted.

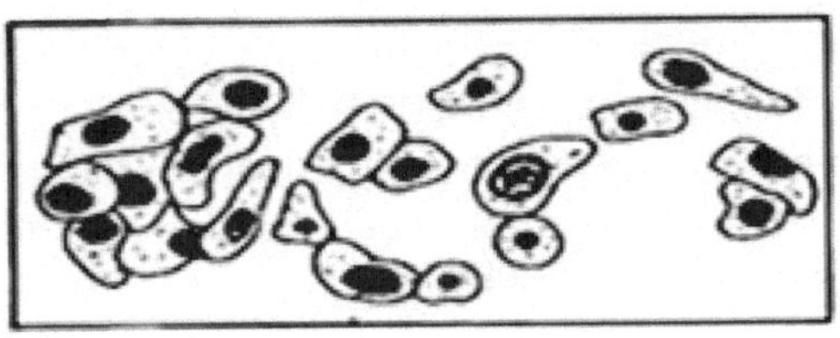

Figure 10.1: Proestrus Phase with Large nucleated epithelial cells.

2. *Estrus Phase*

It is also called heat phase. It is predominant with large numerous cornified squamous cells (keratinized cells). The uteri become highly contractile. In this phase ovulation occurs. It lasts for about 30 hours. Mating is allowed only at this phase. It is the breeding period of rat. In this large amount FSH is secreted

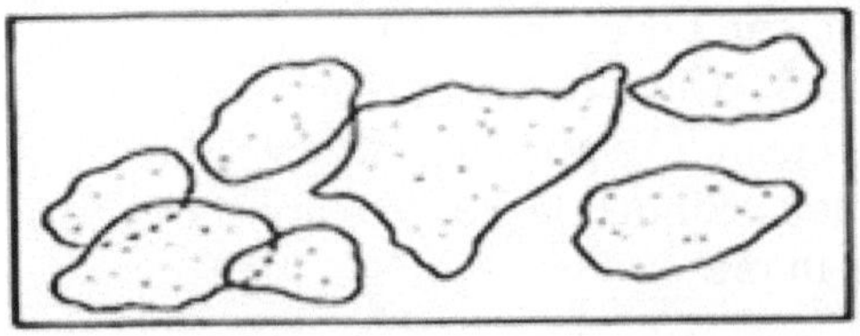

Figure 10.2: Estrus Phase with Cornified Cells.

3. *Metestrus Phase*

It is also called luteal phase. It is characterized by equal number of cornified squamous cells and small rounded numerous leucocytes. It lasts for about 6 hours. It occurs shortly after ovulation. In this phase mating is not permitted. Uteri have diminished in vascularity and contractility. Corpus luteum is well developed.

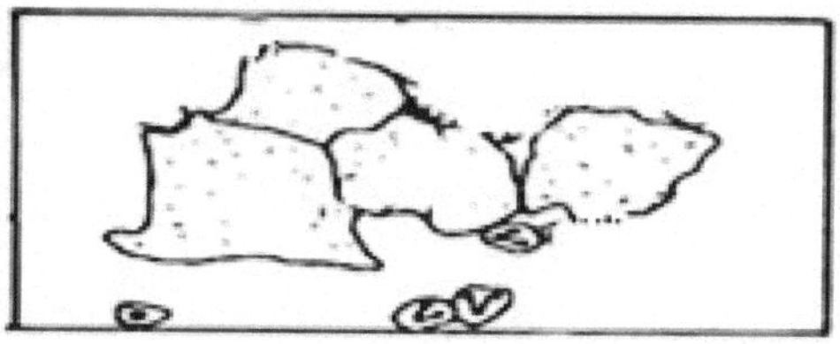

Figure 10.3: Metestrus Phase with Equal Number of Cornified Cells and Leucocytes.

4. *Diestrus Phase*

It is also called anestrus phase. It is predominant with small rounded numerous leucocytes. It is the nonbreeding period of rat. It lasts for about 60 hours. Rgression of corpus luteum occurs.

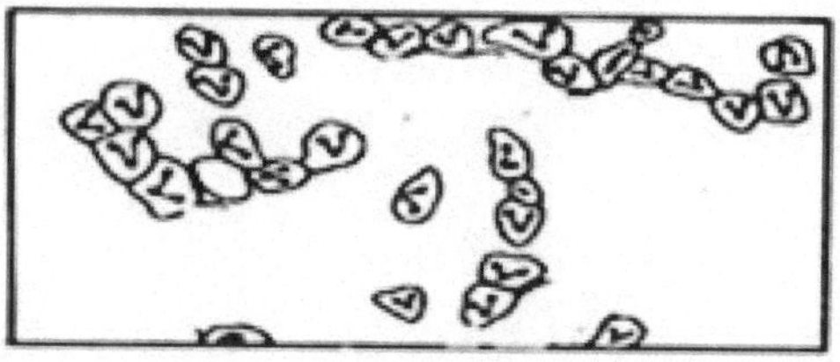

Figure 10.4. Diestrus Phase with Predominant of Small Rounded Leucocytes.

Result

Estrus cycle of rat is completed within 5 days with proestrus, estrus, metestrus and diestrus phases.

Chapter

11

History of Reproductive Organs in Rat

Historical Structure of Reproductive Organs in Male and Female Rat

Aim: To study the histology of reproductive organs in male and female rats.

11.1. T. S. of Testis of Rat

Rat is well known experimental model. It has a pair of white colored, ovoid male gonads called testes. Testes are located in the scrotum. The fibrous muscular part containing the testis and scrotal sac is called gubernaculums. Testis is attached to the dorsal body wall by connective tissue called mesorchium. It is 5 mm in width and 8.5 mm in long. Rat testes measures about 3.7 gm weights. Histological structure includes serosa and tunica albuginea layers, connective tissue, arteries, veins, lacteals, interstitial cells (Leydig cells) and seminiferous tubules. Internal structure of testis includes many septa and lobules. Single lobule contains about 1-3 seminiferous tubules. Retetestis is the location in between long seminiferous tubules and epididymis. In human single testis is formed by about nine hundred seminiferous tubules. These are blind tubules. These are lined by germinal epithelial cells and sertoli cells. In man, it measures about 0.5 meter in long. Sertoli (1865) has been first time reported sperm mother cells or nurse cells in the testis. Germinal epithelial cells produce male primordial germ cells called spermatogonia. Sertoli cell provides nourishment to the spermatazoa and secretes inhibin hormone (X' Hormone). McCullagh (1932) has been first time attempted inhibin hormone from mammalian testis. It acts as macrophages. Blood-testis barrier is due to the Sertoli cells. Spermatogonia under the influence of FSH develop into primary spermatocy (2n), secondary spermatocyte (n), spermatid (n) and sperm (n). The process of formation of spermatid from spermatogonium is called spermatogenesis. The process of formation of sperm from spermatid is called spermiogenesis. The process of release of sperm from sertoli cell is called spermatiation. Both rat testes produce 86 $\times 10^6$ sperms per day. There are specialized endocrine Leydig cells appear to transform from an "undifferentiated" interstitial cell just before the breeding season. Leydig (1850) has been first time studied Leydig cells in testis of mammals. Leydig cells synthesize and secrete testosterone under the influence of LH (Luteinizing hormone) in male called interstitial cell stimulating hormone (ICSH).

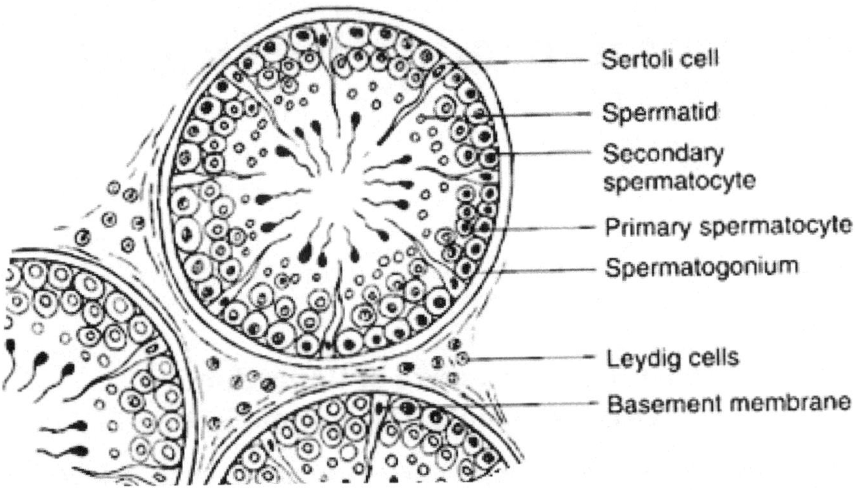

Figure 11.1: T. S. of Testis of Rat.

Function of Testes

1. The production of male gametes called sperms.
2. The synthesis and secretion of testosterone.
3. Testosterone is used in the development of secondary sexual organs and secondary sexual characters like changes in behavior associated with courtship, mating and parental concern.
4. Sertoli cells provide nutrition to the sperms and secretes inhibin hormone.

11.2. T. S. of Epididymis of Rat

It is a coiled, long tubular part located around the testis. In case of human it measures about 6 meters in length. Testis attached with epididymis is called testicle. Histological structure included ciliated columanar cells (stereocilia), basal cells and connective tissue. Epididymis is further divided into 3 parts such as

a. Caputa epididymis: It lies at the top of each testis. It is also called head of epididymis. It collects sperms from retetestis through vasa efferentia.

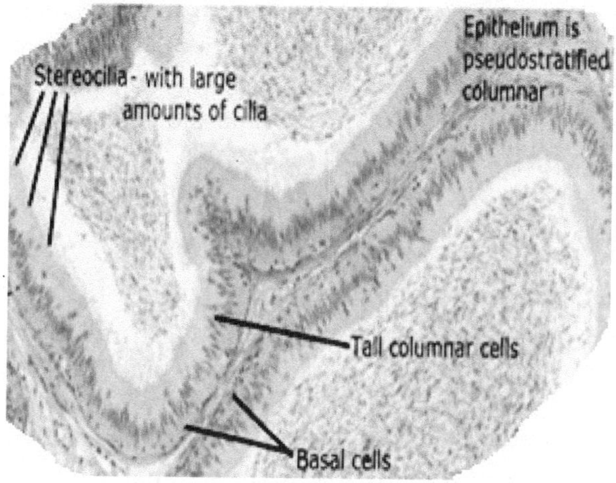

Figure 11.2. T. S. of Epididymis.

b. Corpus epididymis: It is present at the middle of testis and also called body of epididymis. It collects sperms from caputa epididymis.
c. Cauda epididymis: It is posterior part of epididymis and sends a narrow tubular duct called vasdeferens.

Function

1. Storage of sperms.
2. It prevents motility of sperms and maintains fertility.
3. It secretes many inhibitory protein substances.

11.3. A Pair of Seminal Vesicles

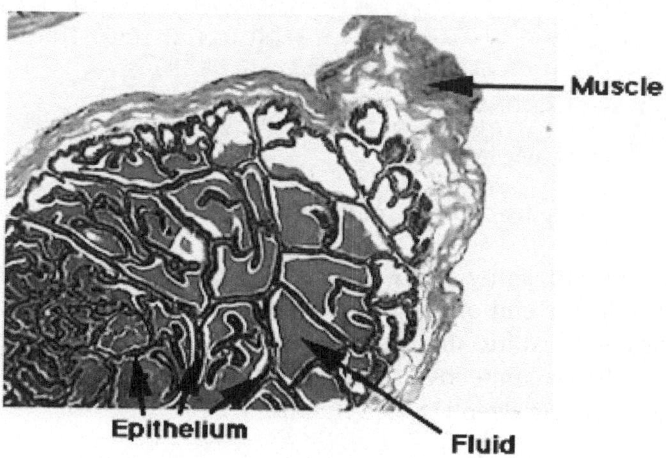

Figure 11.3. T. S. of Seminal Vesicle.

A pair of curved seminal vesicles open into the urethra. These are finger like sacculated glands. Seminal vesicle structurally formed by mucosa layer, muscularis layer (inner circular and outer longitudinal muscle layers) and serosa layer. Mucosa shows secreting epithelium, basement membrane and lamina propria which is loose connective layer rich in arteries, veins and lacteals. The protective outer layer is called Serosa. Seminal vesicle contains stratified epithelium.

Function Seminal Vesicles

i. Secretion of viscous alkaline seminal fluid.

ii. The seminal fluid is yellowish, viscous fluid containing globulin, fibrinogen, citric acid and fructose for the nutrition of sperms. Seminal vesicle does not store sperms. Prostaglandin acts as local hormone and takes part in the transport of sperms towards the site of fertilization. Glucose is converts into fructose in the seminal vesicle and prostate gland (Bhaske, 2018). Prostaglandin action on the cervical mucus enables the easy transport of sperms towards oviduct.

iii. It secretes prostaglandin substances that stimulate uterine contraction during sexual excitation and helps in the transport of sperms towards the site of fertilization. It causes reverse peristaltic movement in the uterus for transport of sperms towards the site of fertilization.

11.4. A Pair of Prostate Glands

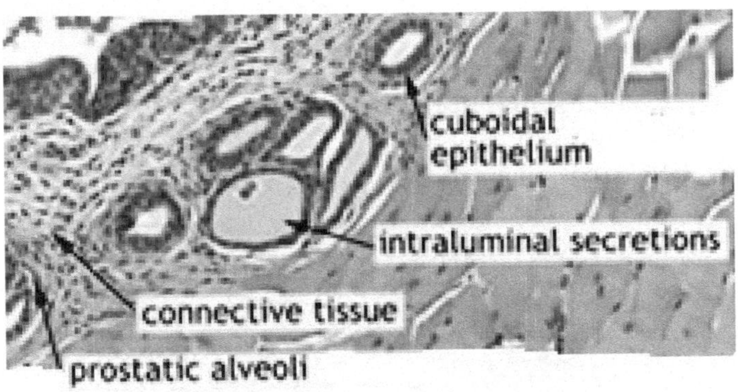

Figure 11.4: T.S. of Prostate Gland of Rat.

These are smooth, shiny, white colored glands about the size of pea. They open into the anterior end of urethra. Structurally, each gland is covered by a fibro-elastic tissue, prostatic sinuses, smooth muscle fibres and partions or and septa containing loose connective tissue called stroma. It has many lobules that are formed by glandular alveoli.

Function of Prostate Glands

i. The alkaline secretion of thin, milky fluid which contains phosphate, calcium, citrate ions and clotting enzyme called profibrinolysin. It's secretion minimizes the acidity of uterus and keeps the sperms active and motile. Prostaglandins help in the transport of sperms towards the site of fertilization. The semen contains 10 per cent sperms, 60 per cent seminal fluid from seminal vesicle and 30 per cent seminal fluid from prostate gland (Bhaske, 2018).

11.5. A Pair of Cowper's Glands

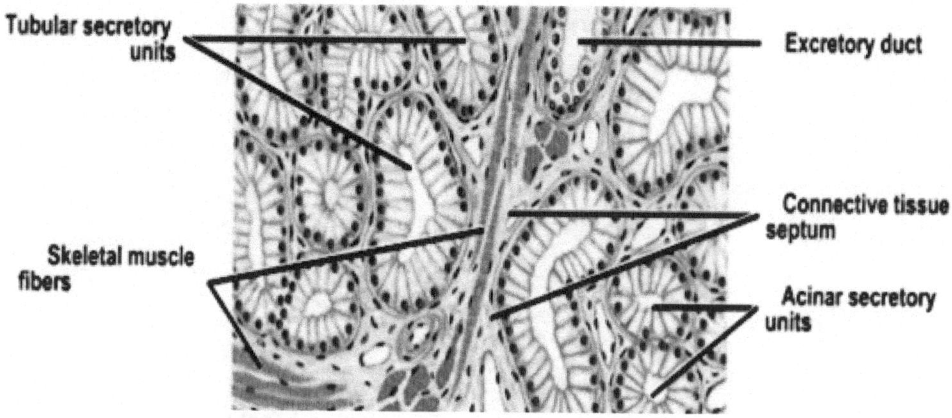

Figure 11.5: T.S. of Cowper's Gland of Rat.

These are small flaske shaped, white colored and measure about 3 mm in diameter. The open into the urethra. It is also called bulbourethral glands. Structurally, each cowper's gland is formed by skeletal muscle fibres, connective tissue, fibroelastic tissues, alveoli (acini) and short ducts that enter into the urethra. Each alveolus has glandular columnar basal nucleated cells and mucous cells. Small ducts unite to form large duct and opens into the urethra. Cowper's glands at the posterior sides show a pair of preputial glands around the opening of urethra.

Function

i. Secretion of an alkaline and clear fluid that in lubrication and neutralizes the acidity of female reproductive tract.

11.6. T. S. of Ovary of Rat

In female rat maturity occurs in between 30 to 50 days of age. Gestation period of rat is 21-30 days. Rat has 5-6 litters (deliveries) per year. Average number of young in a litter is about 8. A pair of ovaries is grape-like, small, pinkish, rounded female gonads lie ventrolateral to the kidneys. Each ovary at the hilum region and suspended by peritoneum called mesovarium. The mesovarium completely encloses each ovary and oviduct within a single compartment called ovarian bursa. Structurally, ovary is covered by a layer of simple cuboidal epithelium called germinal epithelium. The fibrous connective tissue layer present just below germinal epithelium is called tunica albuginea. Each ovary is divided into medulla and cortex.

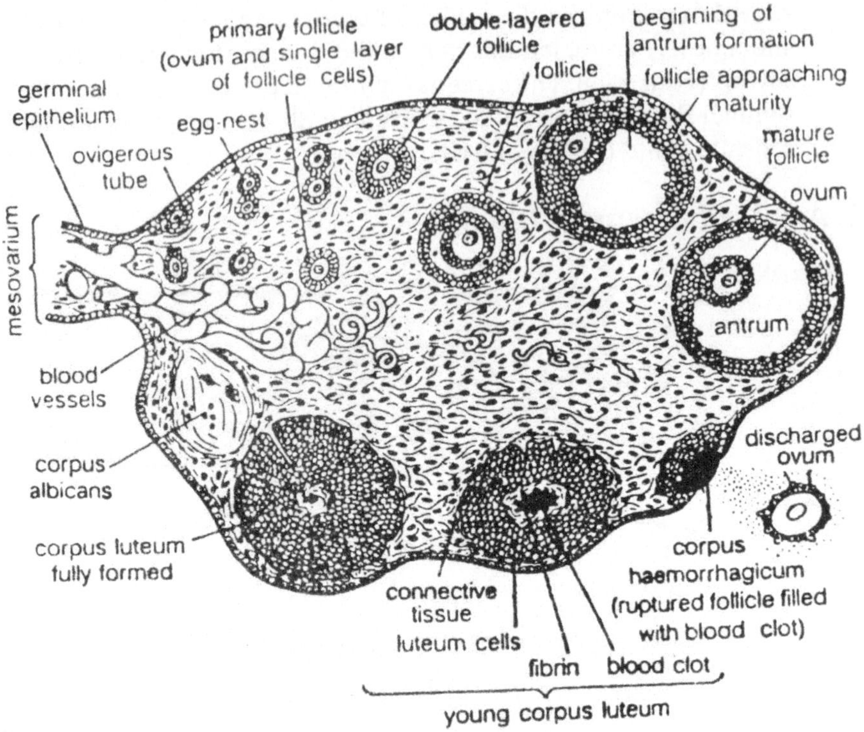

Figure 11.6. T. S. of Ovary.

1. Cortex

It is the peripheral part of ovary. It contains fibrous connective tissue, blood vessels, smooth muscle fibres, lymphatic vessels, nerve fibres, primordial germ cells, mature follicle and corpus luteum. Ovary is lined by germinal epithelium. Germinal epithelium produces primordial germ cells. Teteovarii is present in rodents just above the tunica albuginea layer. The cortex contains numerous primordial germ cells. Primordial germ cells under the influence of FSH develop into primary oocyte (2n), secondary oocyte (n) and mature follicle (n). The cavity of mature follicle (Graafian follicle) is called antrum which is filled with fluid called liquir folliculi. The reformed rupture follicle is called corpus luteum which secretes progesterone hormone. The yellow cytoplasm of luteal cells of corpus luteum is called lutein. The corpus luteum is converted into the corpus albicans and then into the scar. The unmatured primordial germ cells get degenerated are called atretic follicles (atresia). The mammalian ovum was actually discovered by Baer (1827). FSH takes part in the formation of mature follicle. Primordial germ cells divide mitotically to form oogonium. Each oogonium (2n) under the influence of FSH develops into the ovum (n) contains haploid nucleus, mitochondria, Golgi complex and endoplasmic reticulum. Each mature follicle is formed by different layers such as theca externa (fibrous), theca interna (vascular), membrana granulosa (peripheral layers of granulosa cell), discuss proligerous (many layered), corona radiata and zona pellucida. Theca interna layer of mature follicle is also called thecal gland which secretes estrogen. The cavity of Graafian follicle is called antrum which is filled with fluid called liquir folliculi. Liquior folliculi (antral fluid) is rich in hyaluronic acid. The process of release of ovum from ovary is called as ovulation. Ovulation occurs with the influence of LH. Corpus luteum (Latin, corpus = body, luteum = yellow) is present in the cortex of ovary. Reformed ruptured mature follicle is called corpus luteum. The cells of corpus luteum secrete progesterone, small amount of estrogen, inhibin and relaxin hormones. Progesterone helps in implantation, development of embryo and maintenance of pregnancy. If no fertilization then regression of corpus luteum occurs. The degenerating corpus luteum is called corpus albicans and then a small scar of connective tissue.

2. Medulla

Ii is spongy inner part (core) of ovary. It contains loose connective tissue (stroma), blood vessels, spindle shaped fibroblast like cell, elastic collagen fibres, ground substance, lymphatic vessels and nerve fibres. It is devoid of ovarian follicles (Deshmukh, 2003 and Shinde *et al.*, 2009).

Function of Ovary

i. Production of ova.

ii. Secretion of estrogen and progesterone hormones.

iii. Estrogen helps in the appearance of secondary sexual characteristics (mating, parental care) in rat. Estrogen helps in the development of secondary sexual organs like ovary, reproductive tract and mammary glands.

iv. Estrogen helps in the development of mammary glands (six pairs of teats are present in rat).

v. It helps in endometrial changes during estrus cycle. It proliferates epithelial cells of endometrium, increases resistance against infection and helps in the formation of endometrial gland and blood vessels. It changes the vaginal epithelium from a cuboidal into a stratified epithelium.

vi. It helps in bone metabolism. It prevents osteoporosis.

vii. Progesterone helps in the implantation and development of embryo.

11.7. T. S. of Uterus of Rat

There are two uteri in rat. The uterus of this type is called "uterus duplex". Each uterus is called uterine horn or cornu. They are hollow, muscular, vascular organs located in the pelvis region in between the urinary bladder and rectum. Anatomically there are right uterus and left uterus which unite at the terminal region to farm vagina. Structurally, the wall of uterus is formed by three layers such as perimetrium (serosa), myometrium and endometrium (Rastogi, 1998, Deshmukh, 2003, Shidne *et al.*, 2009 and Mutakekar *et al.*, 2011).

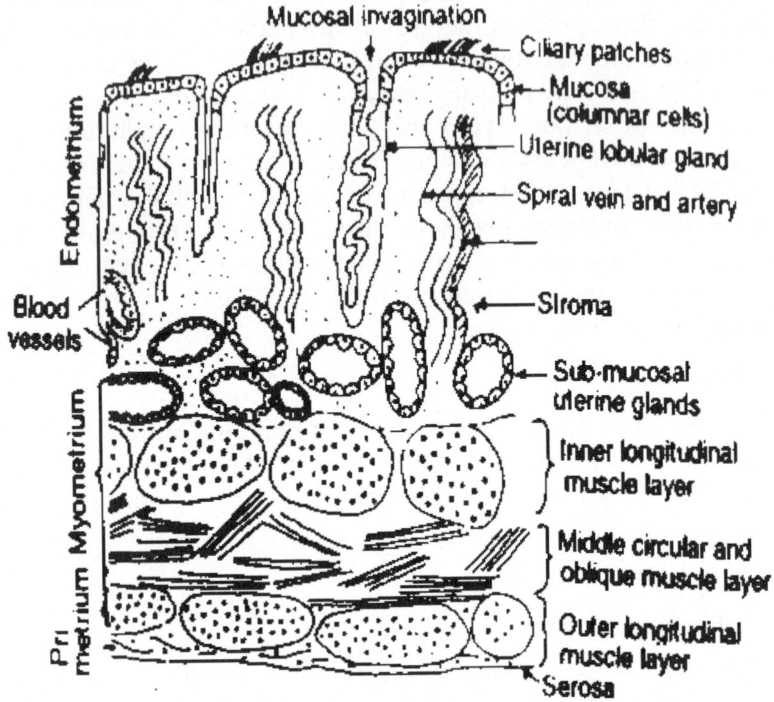

Figure 11.7: T. S. of Uterus of Rat.

i. ***Perimetrium (serosa)***

It is the outer most protective layer (coat) of uterus. It is formed by connective tissue, blood vessels, lymph vessels and nerves.

ii. ***Myometrium or Muscularis***

It is the middle layer of uterus. It consists of bundles of smooth muscles. It is formed by three layers. It's contraction is controlled by oxytocin hormone.

a. ***Stratum Subvascular***

It is inner most layer of myometrium. It is formed by longitudinal muscle fibres.

b. ***Stratum Vascular***

It is middle layer of myometrium. It is formed by circular muscle fibres.

c. ***Stratum Supravascular***

It is outer most layer of myometrium. It is formed by longitudinal muscle fibres.

iii. Endometrium or Mucosa

It is ciliated inner most layer of uterus. It is a mucous membrane consisting of a columnar epithelium and loose connective tissue layer of lamina propria. It is divided into stratum (endometrium) functionalis and stratum (endometrium) basalis. Stratum functionalis periodically shed during estrus cycle and restored. Stratum basalis remains unchanged. Stratum functionalis is the site of embryo implantation. It contains numerous uterine glands. Uterine gland secretes mucin and glycogen called uterine milk used in the nourishment of early embryo.

Function of Uterus

1. It is the site of implantation and development of embryo.
2. It is the target organ for oxytocin hormone
3. It expels the fetus at the end of pregnancy.

11.8. T. S. of Fallopian Tube of Rat

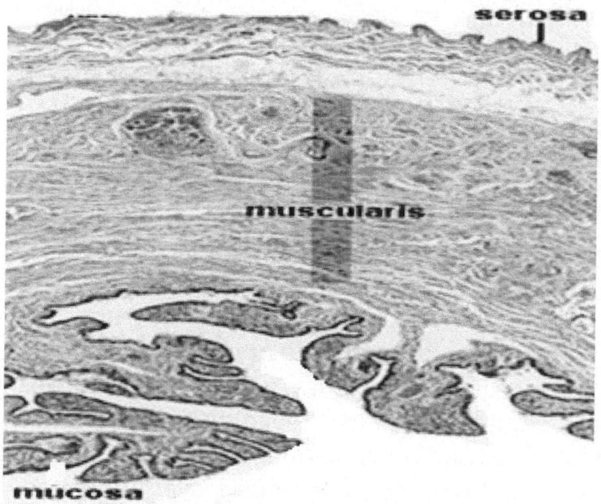

Figure 11.8: T.S. of Fallopian Tube of Rat.

A pair of fallopian tubes (oviducts) opens through funnel. It is about 205 mm in length when coiled. It is formed by mucosa, muscularis layer (inner cicular muscle layer and outer longitudinal muscle layer) and serosa. The mucosa shows transverse folds and ciliated epithelium, lamina propria layer and basement membrane. Lamina propria layer is loose connective layer rich in blood vessels and lacteals. Serosa is outer protective layer. Muscle contraction and sweeping movement in ciliated epithelial cells occur in the fallopian tube at the time of ovulation. Motility of fallopian tubes increased by estrogen and depressed by progesterone. Estrogen increases the ciliated epithelial cells in oviduct. Funnel creates fluid current and receives the ova. Epithelial cells of ovciduct secrete mucus for the easy transport of ovum. It has small opening called infundibulum or ostium.

Function

1. Ostium receives ova from the ovary. Oviduct is the site fertilization.

11.9. T. S. of Vagina

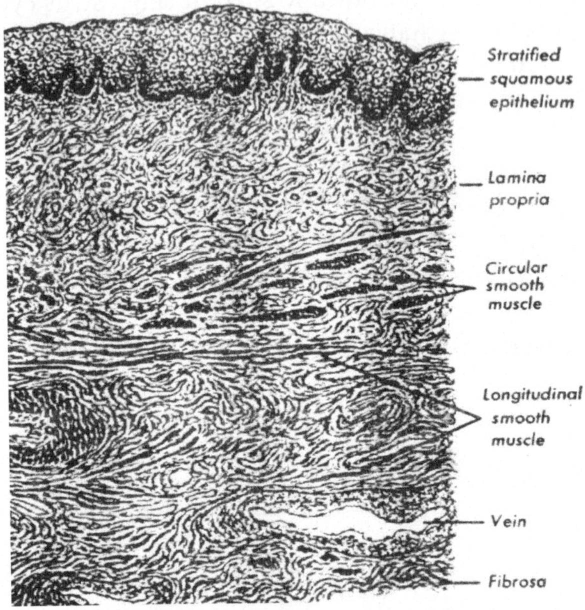

Figure 11.9: T. S. of Vagina of Rat.

Vagina is organ of female reproductive system. Structurally, it is formed by 3 layers such as inner mucosa layer, middle muscularis layer (inner cicular muscle layer and outer longitudinal muscle layer) and outer fibrosa layer (adventitia layer). The mucosa layer shows transverse folds or rugae. Mucosa layer is formed by stratum corneum, stratum mucification, stratum granulosum, lamina propria layer and basement membrane. Lamina propria layer is loose connective layer rich in blood vessels and lymph nodules. The mucosa contains stratified squamous epithelial layer and lamina propria layer. There are about 10 layers of cuboidal stratified epithelium. The superficial mucosal layer shows cornification in estrus cycle. Estrogen increases resistance against infection in the vaginal epithelium. In case of rat urethra opens outside by female urinary aperture and vagina opens by female genital aperture. The vagina and urethra together form a short common passage called the vestibule. A pair of preputial glands as in male also present on either sides of vagina in female rat.

Function

1. It receives sperms during mating.
2. It serves as a birth canal.

Chapter 12

Surgical Techniques

Surgical Techniques in Rat

12.1. Ovarectomy in Rat

Aim: To study the ovarectomy in rat.

Requirement

Mature female rat, tray, anesthetic ether, sterile needle, thread, 70 per cent alcohol, glucose solution, isotonic saline, savalon, dropper, cotton, glass jar, scissor and forceps.

Procedure

1. Rat is placed in a glass jar. Cotton is soaked with ether or chloroform and placed in the jar. The jar is covered with a glass lid. Within a short period rat becomes unconscious. Such rat is used for the experiment. Place the rat on a dissecting tray keeping the ventral surface upward.
2. Remove hairs by sharp razor from anterior abdomen.
3. Wipe the skin surface with 70 per cent alcohol followed by 2 per cent chlorhexidine solution or povidone-iodine solution.
4. Take longitudinal incision of epidermis for about 3 cm in length then gently cut open the dermis.
5. The skin is gently separated from the underlying muscle using cotton swabs or a blunt probe.
6. Locate ovary by visualizing a white spot under the muscle. This white spot is the fat pad covering the ovary.

7. Make a small incision, approximately 0.5cm in length, directly over one of the white spots (ovary).
8. Grasp the fat pad using fine forceps.
9. Soak with sterile isotonic saline.
10. Ligature is applied near to the hilum of ovary to prevent bleeding.
11. Then cut is taken between ligature and ovary.
12. Ovary is then taken in watch glass.
13. First dermis is stitch with two or more stitches and then epidermis.
14. Clean with savalon solution and glucose solution is given to the operated rat.

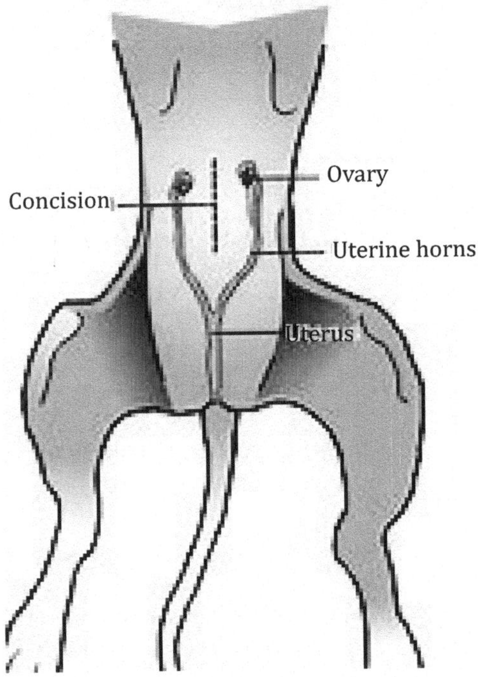

Figure 12.1: Ovarectomy in Rat.

Result

Rat and ovary kept for observation and examination.

12.2. Hysterectomy in Rat

Aim: To study the hysterectomy in rat.

Requirement

Mature female rat, tray, anesthetic ether, sterile needle, thread, 70 per cent alcohol, glucose solution, isotonic saline, savalon, dropper, cotton, jar and scissor.

Procedure

1. Rat is placed in a glass jar. Cotton is soaked with ether or chloroform and placed in the jar. The jar is covered with a glass lid. Within a short period rat becomes unconscious. Place the rat on a dissecting tray.
2. Remove hairs by sharp razor from anterior abdomen.
3. Wipe the skin surface with 70 per cent alcohol followed by 2 per cent chlorhexidine solution or povidone-iodine solution. Take longitudinal incision of epidermis for about 3 cm in length then gently cut open the dermis.
4. The skin is gently separated from the underlying muscle using cotton swabs or a blunt probe. Locate ovary by visualizing a white spot under the muscle. This white spot is the fat pad covering the ovary.
5. Make a small incision, approximately 0.5cm in length, directly over one of the white spots (ovary). Soak with sterile isotonic saline.
6. Ligature is applied near to the top of cervix to prevent bleeding.
7. Then cut is taken between ligature and uterus. Uterus is taken in watch glass.
8. First dermis is stitch with two or more stitches and then epidermis.
9. Clean with savalon solution and glucose solution is given to the operated rat.

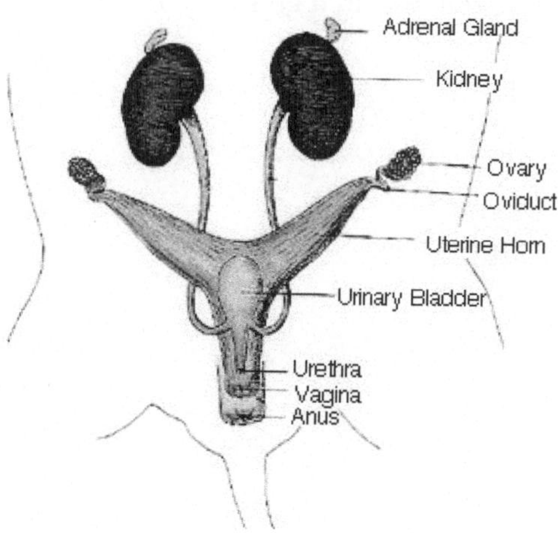

Figure 12.2: Hysterectomy in Rat.

Result

Rat and uterus kept for observation and examination.

12.3. Castration in Rat

Aim: To study the castration (testectomy or orchiectomy) in rat.

Requirement

Mature male rat, tray, anesthetic ether, sterile needle, thread, 70 per cent alcohol, glucose solution, isotonic saline, savalon, dropper, cotton, glass jar, scissor and forceps.

Procedure

1. Rat is placed in a glass jar. Cotton is soaked with ether or chloroform and placed in the jar. The jar is covered with a glass lid. Place the rat on a dissecting tray.
2. Remove hairs by sharp razor from anterior abdomen.
3. Wipe the skin surface with 70 per cent alcohol followed by 2 per cent chlorhexidine solution or povidone-iodine solution. Take longitudinal incision of epidermis for about 3 cm in length then gently cut open the dermis. Incision may be taken at the midventral line of scrotum for this operation. The skin is gently separated from the underlying muscle using cotton swabs or a blunt probe. Locate testis by pressing the scrotum. Ligature is applied near to the base of vas deference to prevent bleeding. Then cut is taken between ligature and testis.
4. Testis is then taken in watch glass.

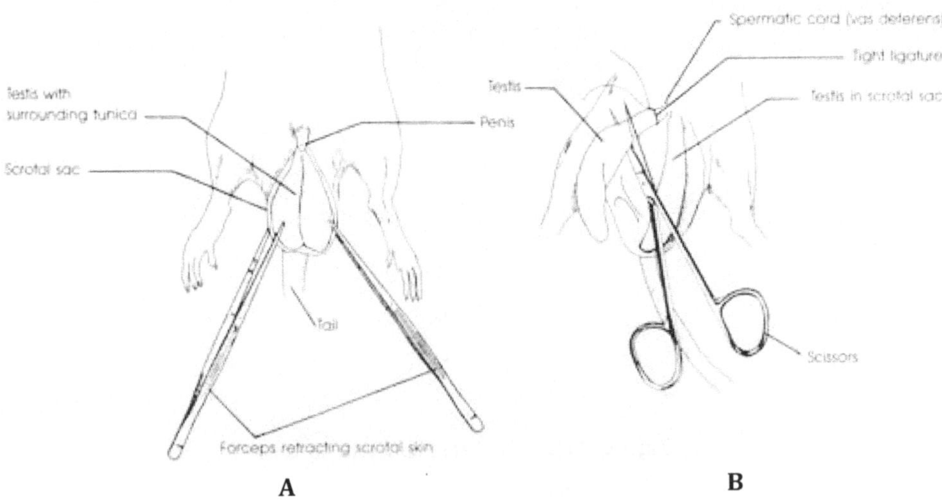

Figure 12.3: Castration in Rat.
A: Exposing the testis; B: Removing the testis.

Result

Rat and testis is kept for observation and examination.

12.4. Vasectomy in Rat

Aim: To study the vasectomy in rat.

Requirement

Mature male rat, tray, anesthetic ether, sterile needle, thread, 70 per cent alcohol, glucose solution, isotonic saline, savalon, dropper, cotton, glass jar, scissor and forceps.

Procedure

1. Rat is placed in a glass jar. Cotton is soaked with ether or chloroform and placed in the jar. The jar is covered with a glass lid. Within a short period rat becomes unconscious. Such rat is used for the experiment.
2. Remove hairs by sharp razor from abdomen.
3. Wipe the skin surface with 70 per cent alcohol followed by 2 per cent chlorhexidine solution or povidone-iodine solution. Take longitudinal incision of epidermis for about 3 cm in length then gently cut open the dermis.
4. The skin is gently separated from the underlying muscle using cotton swabs or a blunt probe. Locate vas deferens. Soak with sterile isotonic saline.
5. Two ligatures are applied to the vas deference keeping suitable distance to prevent bleeding. Then cut is taken between two ligatures and small piece of vas deference is removed. Repeat procedure to the opposite side of vas deference.

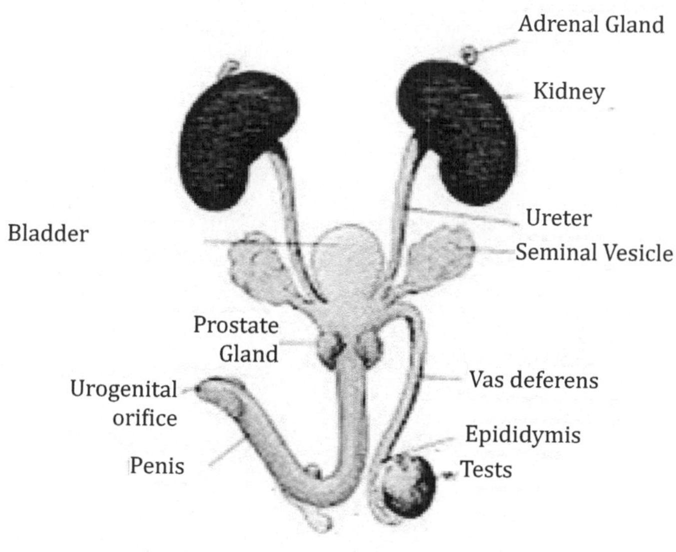

Fig: 12.4: Vasectomy in rat.

6. Small piece of vas deference is then taken in watch glass.
7. First dermis is stitch with two or more stitches and then epidermis.
8. Clean with savalon solution and glucose solution is given to the operated rat.

Result

Rat and small piece of vas deference is kept for observation and examination.

Chapter 13

Study of Animal House

Animal House of Normal and Experimental Animals

Aim: Study of animal house: Set up and maintenance of animal house; breeding techniques; care of normal and experimental animals.

Principle

Experimental animals generally vertebrate and invertebrate animals used in research, testing or teaching. Rodents are generally used as an experimental animals of which mice and rats have been mostly used in laboratories (approximately 90 per cent). The other experimental animals are guinea pigs, pigs, rabbits, cats, monkey, hamster, horse and dogs.

1. Set up and Maintenance of Animal House

1.1. Set up of animal house.

1. Animal house should set up with the permission of the Committee for the Purpose of Control and Supervision of Experiments on Animals (CPCSEA, 1977), India and Institutional Animal Ethics Committee (IAEC). CPCSEA is working under the PCA Act, 1960. Animal welfare board of India is working under the guidelines of PCA Act, 1960. PCA Act, 1960 means prevention of cruelty to animals Act, 1660. Ministry of CPCSEA was started on 08-02-1991.

   The above committee has control over unnecessary pain or suffering on animals.

Table 13.1: Sections of Animal House

STORAGE		ANIMAL HOLDING ROOM
FURNITURE AND EQUIPMENTS	CORRIDOR	PROCEDURE ROOM
WASTE AREA		WASHING ROOM

2. Animal house should set up with the different sections like animal holding section, procedure sections, washing section, waste disposal section, furniture and equipment section, storage section and corridor section.

2. Maintenance of Animal House

1. Animal house flooring should be smooth, moisture proof and non-slippery.
2. Animal house room doors and windows should be dust proof and fit properly within the frames. Rodent barrier can be provided in the doors.
3. Animal house walls and ceilings should be free of cracks.
4. Animal house should be provided with pure water, good humidity, power and light, noise free environment.
5. Animals should have sufficient space to turn around and to express normal postural adjustments.
6. Animals should have ready access to food and water and it must have enough clean area in which to move and rest.
7. A cage is a permanently fixed or movable container that is enclosed by solid walls or wire mesh in which one or more animals are kept or transported.
8. A pen is an area enclosed by walls, bars or meshed wire, in which one or more animals are kept.
9. Rabbits are kept individually in cages with a solid or meshed-wire bottom.

10. Cats are kept in individual caging or in group pens or cages.
11. Dog and swine (miniature pig) may be kept in indoor or outdoor pen areas with open access to the outside run or in an indoor environment.
12. Dogs can be given area to a run or being moved into another area.
13. Monkeys (non human primates) require resting surfaces or perches.
14. Cattles required regular inspection of the veterinarian.
15. Animals should provide daily palatable and nutrition food. Animals should provide drinking water. Animals should provide dry, non toxic bedding.
16. The temperature range for mice, rats, guinea pigs and miniature pigs should be 20-24 °C while for dogs, cats and rabbits 15-21 °C. The relative humidity (RH) have an adverse effect and be maintained below 70 per cent .
17. Animals may be protected from pathogens. The diseases transmitted from one animal to other animals are called zoonosis. Animal quarantine is the separation of a newly received animal.

3. Breeding Techniques of Animal House

3.1. Animal Breeding

The mating of animals for increasing the animals and for improvement in the quality of different characteristics (size, feeding behaviour) is called animal breeding.

3.2. Methods of Animal Breeding

There are two methods of animals breeding such as inbreeding and out breeding.

i. Inbreeding

Breeding in between animals of the same breed for 4-6 generations is called inbreeding. Inbreeding may be explained by taking an example of cows and bulls. Superior cows and superior bulls of the same breed are identified and mated. The progeny obtained from such mating are evaluated and superior males and females are identified for further mating. A superior female, in the case of cattle, is the cow that produces more milk per lactation. Superior male is that bull, which gives rise to superior progeny as compared to those of other males. Homozygous pure genotypes are developed as per Mendelian inheritance. This helps in restoring fertility and yield.

ii. Out Breeding

Breeding in between animals of the unrelated animals which may be between individuals of the same breed (but having no common ancestors) or between different breeds (cross breeding) or different species (interspecific hybridization).

It is the mating of animals within the same breed but having no common ancestors on either side of their pedigree up to 4-6 generations. The offspring of such a cross is called as an outcross. Out crossing is the best breeding method for milk production, growth rate in cattle *etc.*

iii. Cross-breeding

In cross-breeding superior males of one breed are mated with superior females of another breed. Many new animal breeds have been developed by this strategy. It gives better breeds. Cows of an inferior breed may be mated to bulls of a superior breed to get better progeny. Hisardale is a new breed of sheep developed in Punjab by crossing Bikaneri ewes and Marino rams.

iv. Interspecific Hybridization

In this male and female animals of two different species are mated. The progeny obtained from such a mating are usually different from both the parental species. In some cases, the progeny may combine desirable characters of both the parents. Mule is produced from a cross between female horse (mare) and male donkey. Mules are harder than their parents and are well suited for hard work in mountainous regions.

4. Care of Normal and Experimental Animals

Care of normal and experimental animals should be taken as per the PCA Act 1960 and rules of breeding of and experimentation (control and supervision) Rules 1998 as amended in 2001 and 2006. Experiment means any programme/project with use of animals for the acquisition of scientific knowledge.

1. Every experiment shall be performed under the supervision of person duly qualified in animal science and medical science of recognised institution.
2. Animals should not feel pain in the experiments.
3. Animal death and experiments should be avoided.
4. Small animals like rat and rabbits should be selected for study.
5. Looking after of animals before, during and after the experiments.
6. Provide veterinary care to animals and take care in transportation of animals.
7. Experimentation on living animals may be under the close supervision of qualified and experienced person.
8. Select procedures like isolation cell, organ and tissue culture.
9. The goal of any sanitation program is to maintain sufficiently clean, dry bedding, adequate air quality and clean cage surfaces and accessories (Garber, 2011).

Chapter

14

Human Vaginal Exfoliate Cytology

Aim: To study the human (woman) vaginal exfoliate cytology.

Principle

Vaginal exfoliate cytology is studied from spontaneous shedding of cells from the lining of vagina into a cavity. The samples are derived from several sources. Examples like vaginal smear, urine and cerebrospinal fluid. Growth of the vaginal epithelium occurs during the follicular phase of the menstrual cycle, accompanied by the deposition of glycogen and mucopolysaccharides in this tissue. Follicle secretes estrogen that helps in the growth of vaginal epithelium. Epithelial cell are exfoliated into the vagina and are a reliable index of estrogen action.

Requirements

Glass pipette with rubber suction bulb or plastic disposable bivalve speculate, microscope, 0.9 per cent physiological isotonic saline solution of NaCl, glass slide, cotton, eosin and methylene blue stain.

Procedure (Papanicolaou method)

1. Two drops of saline solution (0.9 per cent) are inserted into the vagina of woman with the help of glass pipette with rubber suction bulb (Papanicolaou method).
2. After 5 minutes vaginal smear is taken by the glass pipette with rubber suction bulb and mounted on the glass slide. Vaginal surface biopsy can also be taken by plastic disposable bivalve speculate (Ayre method). Nuclear stain like methylene blue is used for staining the nucleus. Eosin is used for staining cell cytoplasm. The presence of normal or abnormal epithelial cells are the criteria for the study.

3. Vaginal smear containing squamocolumnar cells are smeared on a clean glass slide and observe under microscope. It is important to that air drying before fixation results in destruction of the cells and the smears are then useless for study.

Figure 14.1: Normal Squamocolumnar Cells in Human Vaginal Exfoliate Cytology.

Result

Human vaginal exfoliate cytology showed normal squamocolumnar cells. Single nucleus in the cell, regular outline, same cell size and shape, normal cytoplasm indicates presence of normal epithelial cells in the vaginal smear.

Clinical Significance

1. It helps in the diagnosis of female endocrine disturbances.
2. It helps in the detection of infection with a high-risk human papillomavirus (HPV) which causes cervical cancer.
3. It also helps in the detection of carcinoma of the uterine cervix by detecting the miRNAs.
4. Cervical cancer is the fourth most common cancer in women, with an estimated 528,000 new cases and 265,000 deaths worldwide in 2012 and approximately 90 per cent of cervical cancer deaths occur in developing countries (Champion, 1967).

Chapter 15

Study of Sperm Count and Sperm Motility

Aim: To study the sperm count from given semen sample.

Principle

Semen is secreted by male reproductive organs. Semen contains sperms and seminal fluid. The process of coagulation of semen occurs due to the prostate gland protein like fibrinogen. The process of liquefaction of coagulated semen occurs due to the prostate gland enzymes within 10 to 20 minutes. The liquefied semen is translucent, turbid and viscous fluid. The liquefied semen after dilution with semen diluting reagent is used for sperm count. Normal volume of semen is 2.5 mL to 4 mL and normal sperm count is 400 million sperms. Normal sperm count is 60 to 150 million sperms per mL. Man is oligospermic with sperm count below 60 million sperms per mL and sterile in 20 million sperms per mL. Normal man has 80 per cent sperms motility. Normal man semen complete liquefaction within 15 minutes.

Requirements

1. Neubauer ruling or Thoma-Zeiss haemocytometer, thick glass coverslip of haemocytometer, compound microscope, W.B.C. diluting pipette, 0.1 mL graduated pipette or Hermann Sahli haemoglobin pipette.
2. Semen diluents reagent for sperm count: Semen diluting reagent or fluid contains 5 gram sodium bicarbonate, 1 mL neutral formalin and 100 mL distilled water.

3. Semen sample: Semen sample should be collected in a clean dry container. Semen sample should be examined within an hour in the laboratory. Before collecting the specimen, the patient should have refrained from ejaculation for at least 2 days.

   Never refrigerate the semen. Never use condom for collection that may contain spermicidal agents and impair sperm mortality. Normal semen sample is a highly viscous, opaque, white colored and with acrid odour. Freshly collected semen undergoes coagulation and after 10 to 20 minutes leads to translucent, turbid and viscous fluid. Semen sample undergoes coagulation and liquefaction.

Procedure

1. Mix the semen sample thoroughly by shaking the container tube gently.
2. Take 10 mL of the diluting reagent and 0.5 mL of liquefied semen sample in glass test tube. This makes 1:20 dilution.

   Alternatively:

   2.1. Take 0.38 mL of the diluting reagent and 0.02 mL of liquefied semen. This makes 1:20 dilution.

   2.2. Take 19 drops of the diluting reagent and 1 drop of liquefied semen sample. This makes 1:20 dilution.

   2.3. Take liquefied semen up to the 0.5 mark of a W.B.C. diluting pipette and then the diluting reagent up to the 11 mark (Ghose and Manna, 2016). Mix well and this makes 1:20 dilution. Note: If blood pipette (sahli) is used, mix 0.95 mL of diluting reagent with 0.05 mL of liquefied semen).
3. Fill a hemocytometer chamber by diluted semen and keep for two minutes. This allows the immobilized sperms to settle down. In hot climate, cover with petri dish in order to prevent from drying. The tip of the pipette is placed in the V-shaped groove on the hemocytometer during the loading of diluted semen into the chamber. Capillary action will draw the fluid into the chamber.
4. Count the number of sperms in the four corner squares covering 4 sq mm under the low power of compound microscope (the microscopic field covers 1 sq mm). While counting, consider only those sperms are complete that is with heads and tails.
5. The full grid on a hemocytometer contains nine squares, each of which is 1 mm square. It is divided into 5×5= 25 small squares. Each this small square again divided into 4×4= 16 smallest squares. The central counting area of the hemocytometer contains 25 large squares and each large square has 16 smaller squares. When there are 200 to 400 sperms per square the sample is at the proper dilution. For determination of sperm count take 4 squares within the large center squares.

The number of sperms in four squares was counted.

Calculate

$$\text{Sperms count per mL} = \frac{\text{Sperms counted x 20 x 1000*}}{4 \times 0.1}$$

For example sperms counted in four squares = 261+337+224+405

For example sperms counted in four squares = 1227

Sperms count per mL = Sperms counted x 50,000

Sperms count per mL = 1227 x 50,000

Sperms count per mL = 61, 350000

Sperms count per mL = $6.1\ 35 \times 10^7$

Note: * 1 mL = 1 cu cm (cc)=1000 cu mm.

Many textbooks call for counting only 2 sq mm (two big squares).

The above factor in this case will be 100000 or sperm count x 100000.

Note: The blood cell count is expressed in cu mm (or µl) whereas the sperm count is expressed per mL (cc).

Result: The given semen sample has 6.135 x 107 sperms/mL.

Clinical Significance

1. It is useful in male infertility and can be corrected by proper medical care.
2. Presence of acid phosphatase in the vaginal fluid is a reliable test for the diagnosis of rape victim. Semen has pH 7.7.

Study of Sperm Motility

Aim: To study the sperm motility in given semen sample.

Principle

Semen is secreted by male reproductive organs. Semen contains sperms and seminal fluid. The liquefied semen after dilution with semen diluting reagent is used for sperm count. Normal man has 80 per cent sperms motility. Normal man semen complete liquefaction within 15 minutes. Sperm motility is attempted by using microscope.

Requirements

Glass slide, glass coverslip, microscope with 45_X objective, graduated 0.1 mL pipette or Hermann Sahli haemoglobin pipette and liquefied semen.

Semen Diluents Reagent for Sperm Motility

Semen diluting reagent or fluid contains 5 gram sodium bicarbonate and 100 mL distilled water. Avoid formalin addition for sperm motility study that kills the sperms.

Procedure

1. Place a drop of liquefied semen on a dry pre-warmed (37 °C) glass slide.
2. Cover the drop with a coverslip by the "wet mounting technique".
3. Rim the coverslip with Vaseline to avoid drying.
4. Examine the slide under the high power objective.
5. Scan several fields until a total of at least 200 sperms have been observed.
6. As sperms are hyaline, cutting down of light helps to locate them.
7. Count the number of actively motile sperms out of the total count of 200.
8. It is essential to focus through the entire depth of a given field so as to include non-motile sperms that have settled to the bottom of the medium.
9. Calculate the percentage of sperm showing actual progressive motion. 10. In this case, roughly note the proportion of motile cells and if most of them (80 per cent) are actively motile, it will be considered as normal.
10. Examine the slide and report the number of motile sperms.
11. The motile forms will decrease by about 5 per cent per hour after the fourth hour following collection.
12. Since sperm must penetrate the cervical mucus within a few minutes following ejaculation, the observation of motility after few hours has some value.

Result

The given semen sample showed more than 80 per cent of sperm motility.

Clinical Significance

1. Less than 80 per cent sperm motility indicates poor sperm motility.
2. Forward straight motility indicates normal sperm and circular motility indicates abnormal sperms. Abnormal sperms may be inactive.
3. Normal sperm motility indicates normal functioning of male genital tract and accessory glands.
4. It is useful in male infertility and can be corrected by proper medical care. Healthy sperms will stay active even after 6 hours but completely cease to be active after 12 hours. Weak sperms cease their activity before 3 hours. In sperm motility forward progression of sperms occurred at least 25 micrometers per second (Ghose and Manna, 2016).

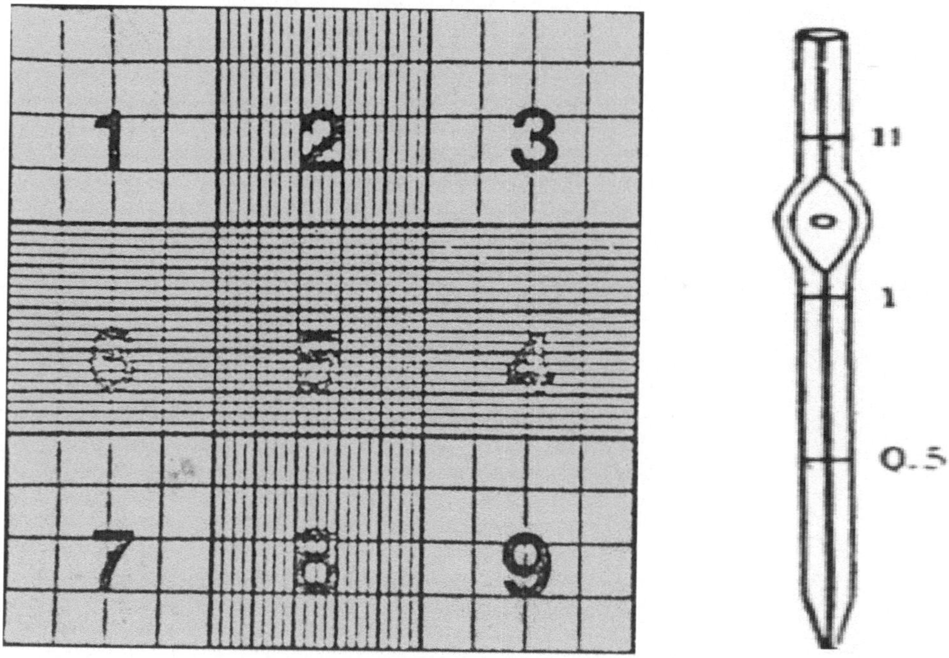

A B

Figure 15.1: A: Neubauer ruling counting area; B: W.B.C. diluting pipette.

Chapter 16

Contraceptive Devices

Modern Contraceptive Devices

Aim: Study of modern contraceptive devices.

Principle

The prevention of pregnancy is called contraception. The devices used for prevention of pregnancy are called contraceptive devices. There are following different methods of contraception.

Modern Contraceptive Techniques

The prevention of pregnancy or conception is called contraception. The devices used for prevention of pregnancy are called contraceptive devices. It is also called control of fertility. The control on population is a big problem before the human community. Fortunately the family planning program is being implemented in almost all countries of the world. The different methods help to prevent unwanted pregnancies included temporary method, permanent method, Medical termination of pregnancy (MTP) and other method.

A. **Temporary Method**

1. ***Natural Methods***

   1.1. **Natural methods/Safe period/Rhythm method:** It includes sexual relation by avoiding possible day of ovulation called safe day. A week before and a week after menstrual bleeding is considered the safe period for sexual intercourse. This idea is based on the fact that ovulation occurs on the 14th day of menstrual cycle. Sexual relation may be avoided in three days before ovulation (11, 12, 13) and three days after ovulation (15, 16, 17).

1.2. Total abstinence of coitus: In this male and female avoid sexual intercourse.

1.3. Coitus interruptus/Withdrawl: It is the sexual relation without deposition of sperms. In this male withdrawls his penis from the vagina just before ejaculation. It has drawbacks that the pre-ejaculation fluid may contains sperms and can cause fertilization.

1.4. Lactation amenorrhea/Absence of menstruation: In lactation period ovulation gets stopped for about six months. In this period there will be no pregnancy. However, it has high chances of failure.

2. *Mechanical Method/Physical Barrier Methods*

It includes male condom and female condom (caps).

2.1. Male Condom (Nirodh)

Figure 16.1: Male Condom.

It is nonporous elastic extensively used of all contraceptive appliances. It is temporary mechanical method of contraception. It creates a barrier between the male organ and interior of female reproductive tract. It is placed over penis during intercourse. It collects the semen. It prevents deposition of sperms in the vagina. It is elastic, rolled and placed in the packet. Condom prevents infection of sexually transmitted disease like AIDS. It prevents sperms passing into the cervix. It is discarded after use. Condom may damage by oil-based lubricants. There are two types of condoms called male and female condoms. A male condom is a sheath that covers the penis. Male condoms are more popular and widely available. Chemically, condom is formed by latex, polyurethane and polyisoprene. Polyisoprene is a synthetic form of latex. There are two types of condoms such as lubricated and unlubricated condoms. Lubricated condom contains spermicidal chemical (chemical like nonoxynol-9) which kills sperm. Most doctors recommend that you use condoms that don't contain spermicide. Air should be removed from the condom and space should be kept for semen. This important step helps to prevent the condom from breaking. Never reuse a condom. Condom is helpful for preventing premature ejaculation in some men. In some cases condom causes low sensation, skin irritation, contact dermatitis and allergy due to chemicals used in the condoms.

Function

1. It prevents the process of pregnancy.

2.2. Female Condom

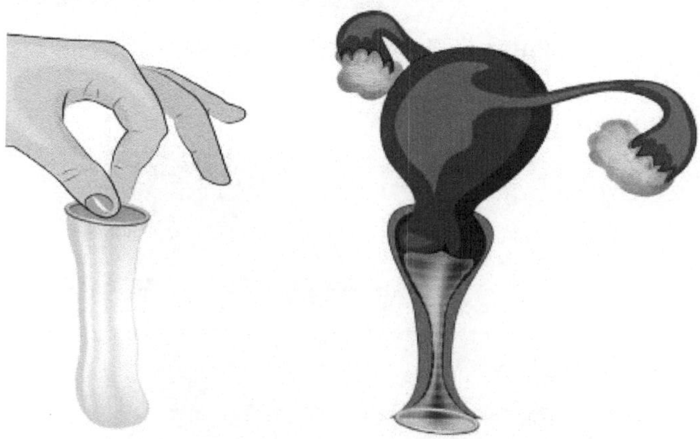

Figure 16.2: Female Condom.

It is nonporous elastic mechanical contraceptive device that prevents sperm from entering the uterus. Condom prevents infection of sexually transmitted disease like AIDS. It prevents sperms passing into the cervix. It is called mechanical method of contraception. Condom may damage by oil-based lubricants. It creates a barrier between the male organ and interior of female reproductive tract. It is inserted into the vagina during intercourse. It collects the semen. It prevents deposition of sperms in the vagina. It is elastic, rolled and placed in the packet.

2.3. Female Cervical Cap

It is nonporous, made of rubber, metal or celluloid, plastic birth control device. It is 4 cm in diameter. It is called as mechanical method of contraception. It is held in place by suction. It creates a barrier between the male organ and interior of female reproductive tract. It is inserted over the cervix by doctor.

Figure 16.3: Female Cervical Cap.

2.4. Female Diaphragm

Figure 16.4: Female Diaphragm.

It is nonporous plastic dome shaped covering with spermicide that kill the sperms. It is reusable flexible rubber or silicone cup like contraceptive device. It creates a barrier between the male organ and interior of female reproductive tract. It is inserted over the cervix by doctor. It is also called Dutch cap. It is inserted at any time within two hours before intercourse and should be removed six to eight hours after intercourse. Plain warm water may follow its withdrawal but is not necessary. With proper care of the diaphragm after use it should be washed with soap and water, thoroughly dried and dusted with talcum powder it should remain in good condition for at least a year. The satisfactory use of diaphragm requires care and intelligence.

Function

It prevents the process of pregnancy.

2.5. Intra Uterine Contraceptive Devices or Long Duration Contraception

It includes intra uterine device (IUD) like Copper-T and Lippes loop. It is an intrauterine contraceptive device (IUCD or ICD). It is inserted into the uterus by doctor. It is called as mechanical method of contraception. It creates a barrier between the male gamete and female gamete. It also does not allow implanting the fertilized egg. It is longer duration reversible contraceptive device (birth control). It prevents the process of fertilization. It can be replaced after three years. It is 100 per cent effective contraceptive device. However, it has a failure rate of about 0.8 per cent. It is painless device. Sometimes, it causes pain, backache, menstrual bleeding.

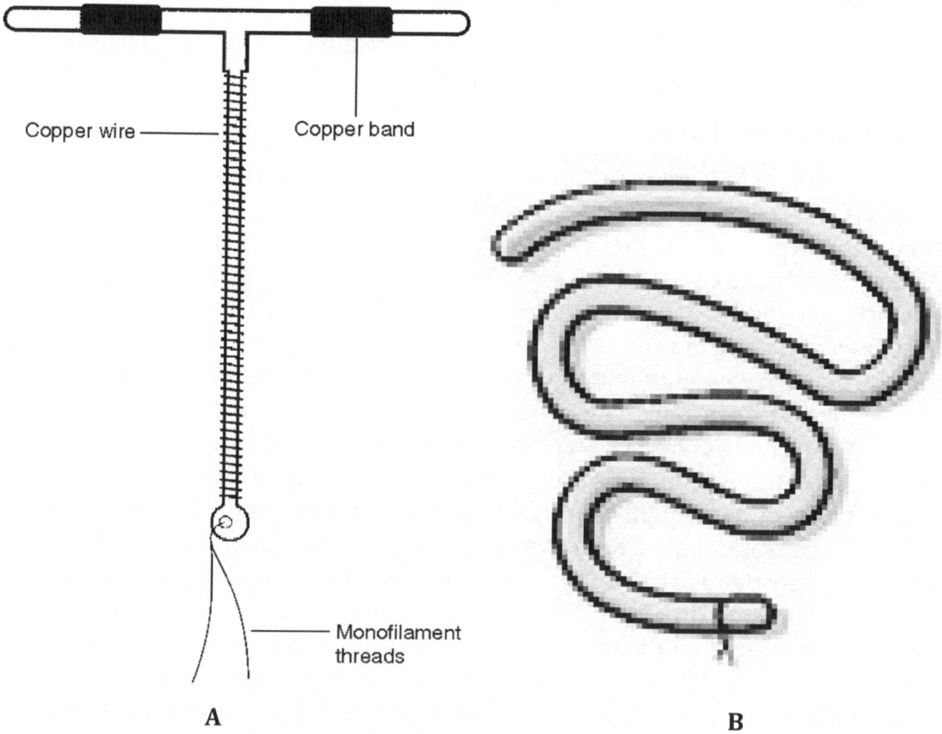

Figure 16.5: Intra Uterine Contraceptive Devices.
A: Copper-T; B: Lippes loop.

The effects are not serious and disappear after few months. Copper-T is metallic made from copper band, copper wire, stainless steel and monofilament threads. Copper loops are plastic double "S" loop and not in commonly used. Copper T is widely used. It is 100 per cent effective contraceptive device.

Function

It prevents the process of pregnancy.

3. Chemical Method/Physiological/Oral Pills

3.1. Foam tablets

It includes foam tablet which produces foam when it is moistened. The foam is spermicidal.

3.2. Jelly

It includes jelly which is inserted into the vagina with help of applicator. The jelly is spermicidal that kill the sperms.

3.3. Cream

It includes cream which is inserted into the vagina. The cream is spermicidal that kill the sperm (NONOXYNOL-9 and TODAY).

3.4. Oral Contraceptives or Hormonal Contraception

Birth control pills taken by female in which daily moderate level of estrogens suppress the ovarian and menstrual cycles. It contraceptive pills like CHOICE tablets; saheli (nonsteroidal); ESME-28; Enovid; Enovid-E; Ortho-novum; Noristrein; Provest; Ovulen and Oracon. It involves oral administration of chemical substances, popularly called pills. Saheli (nonsteroidal) is to be taken once in a week. The pills affect internal condition in such a manner that conception can be terminated. Pills act in such a way that neither ovulation nor implantation will occur. Oral pills are nothing but hormones like estrogen, progesterone orb analogous synthetic steroids. It is called as chemical method of contraception. It is also called hormonal method of contraception. Pills are started from fifth day of menstrual cycle. Pills are generally taking at night time. If pill taking is missed then take one pill at next morning. It causes negative feed regulation on pituitary gland. This is achieved by inhibiting release of FSH and LH from pituitary gland. Secondly, the endometrial surface modify is modify in such a way that it does not allow implantation of the zygote or maintenance of the pregnancy. Thirdly, it makes the cervical mucus impermeable to the sperm and difficult for the sperm to penetrate. Oral pill can be started from 5th day of menstrual cycle. The one oral pill every day preferably at night may be taken. Pills containing estrogen and progesterone hormones create negative feedback on the anterior lobe of pituitary gland and decrease the secretion of gonadotropic hormones (FSH and LH) that prevent the production of ova. The low secretion of estrogen and progesterone prevent ovulation. Sometimes, pills cause gastrointestinal changes, nausea, edema, headache, vomiting, diarrhea, mood changes, breast tenderness or moderate increase in weight or decrease in weight, blood clotting, high blood pressure and irregular bleeding. Progesterone type of birth control pill is called as progestin (minipill). Your doctor can help you decide which pill is right for you. The contraceptive pills work by suppressing pituitary gland so that the release of an egg from ovary or ovulation doesn't occur. The contraceptive pills thickens cervical mucus to prevent sperms from reaching towards ovum. The lining of uterus is also altered to prevent implantation. The missing or skipping doses causes hormone levels to fluctuate. It includes emergency contraception pills. The tablet unwanted 72 hours may be taken with 72 hours for unprotected sexual relation (Nagbhushanam *et al.*, 1996).

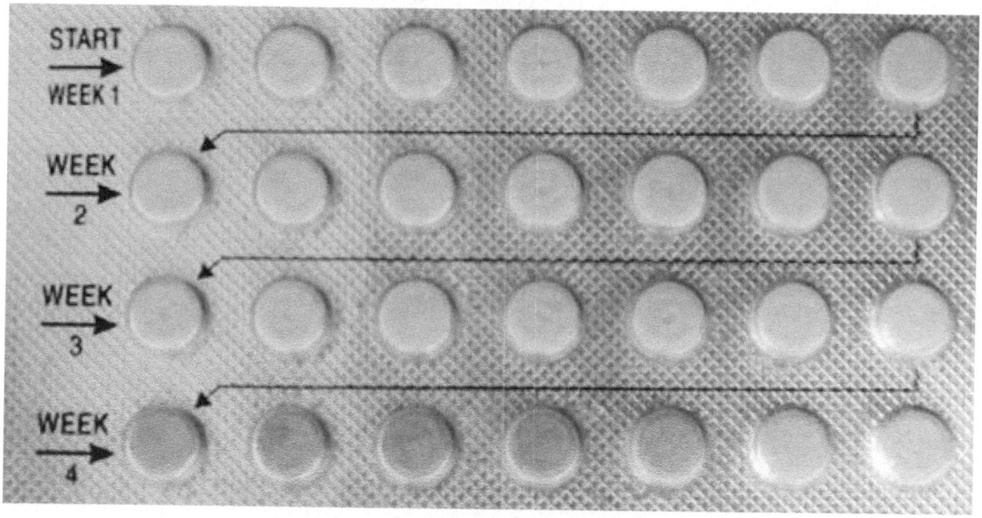

Figure 16.6: Oral Contraceptives.

Function

It prevents the process of pregnancy.

B. Permanent Method

1. *Sterilization Operations*

The permanent birth control method in men is called vasectomy and in women it is called tubectomy. These are surgical permanent method, also called sterilization.

1.1. Sterilization Operations in Male (Vasectomy)

It is a surgical minor operation in male (time required 20 minutes). Vasectomy done by male where the vasa deferentia tubes are cut to prevent sperm transport. In vasectomy longitudinal incision of skin is taken on the scrotum and vasa deferentia are located. Two ligatures are mad for each vas deferens and small piece of vas deference is removed. The procedure is repeated to the opposite side of vas deference. In this passage of sperms is blocked. After operation he would not carry weight for at least one month. This method is 100 per cent effective. It is carried out in primary health centre by doctor.This blocks, gamete transport and prevents pregnancy.

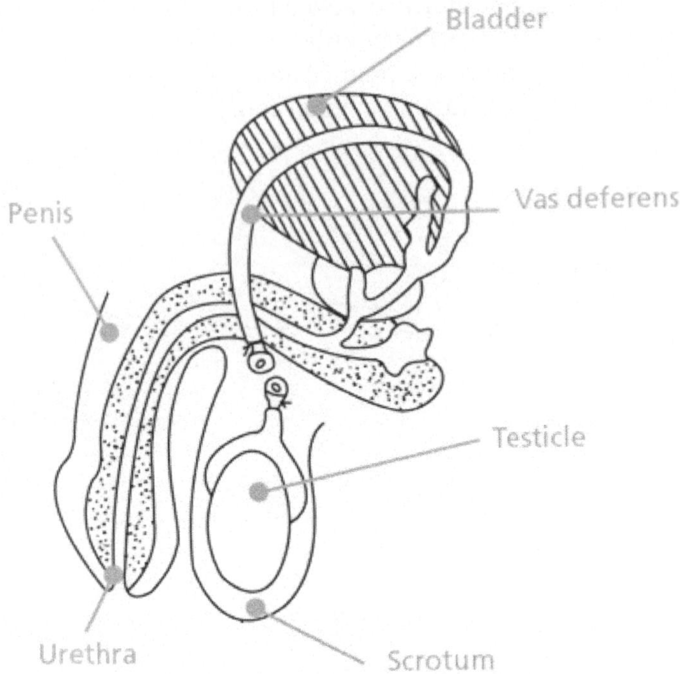

Figure 16.7: Sterilization Operations in Male.

1.2. Sterilization Operations in Female

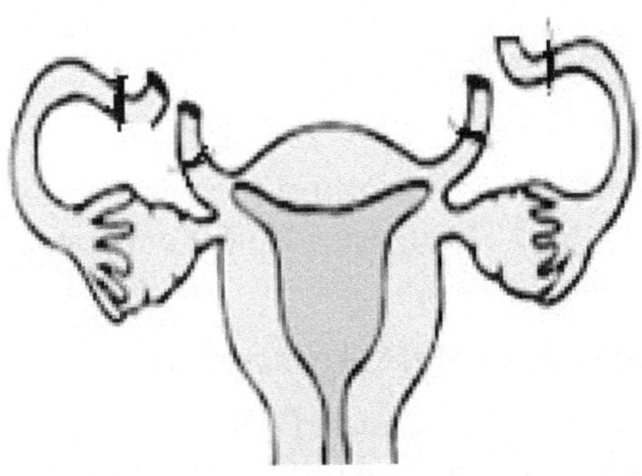

Figure 16.8: Sterilization Operations in Female.

It is a surgical minor operation in female. Tubectomy done by female where the fallopian tubes are tied or cut to prevent ovum transport and passage of sperm. In tubectomy longitudinal incision of skin is taken on the abdomen and fallopian tubes are located. Two ligatures are made on each fallopian tube and small piece of fallopian tube is removed. The procedure is repeated to the opposite side of fallopian tube. In this passage of ovum is blocked. After operation she would not carry weight for at least one month. This method is 100 per cent effective. It is carried out in primary health centre by doctor (Nagbhushanam *et al.*, 1996).

C. Medical Termination of Pregnancy (MTP)/Abortion

It includes voluntary termination of pregnancy through doctor. Prostaglandins appear to be the most promising substance and can be administered by different routes with success. Thus, Prostaglandin E and F are introduced into the vagina in an amount 40-100 mg, they terminate pregnancy of two to seven days as counted from the expected day of menses. It is safe during the first trimester of pregnancy. It is induced abortion and includes medical termination of pregnancy (MTP). Abortion includes dilation and surgical curettage (DC). MTP includes dilation of cervix and sucking of embryo. Stilboestrol (but not natural estrogen) given early in pregnancy interferes with the implantation of the ovum and may induce abortion. Government of India has legalized MTP Act in 1971, with strict conditions to avoid its misuse. Medical termination of pregnancy (Amendment) Act 1971 was enacted by Government of India. (Nagbhushanam *et al.*, 1996; Shine *et al.*, 2011 and Adate and Ajagekar, 2019).

D. Other Method

The birth control implant is a contraceptice uswed by the female. It is a tiny, thin rod about the size of a matchstick. It is implanted under the skin of the upper arm. They contain progesterone and estrogen. Their mode of action is similar to that of pills. They prevent pregnancy for 3-4 years.

Chapter 17

How to Write Educational Tour Report

Zoological Study Tour

Educational tour report is to be written for submission in the University examination with the help of information given below.

1. Date and time of visit to study area.
2. Name of teacher guide and administrative staff.
3. Introduction, name, types (national or state level), administration, *etc.*, weather condition and period of visit of study area.
4. Objectives and importance of visit be mentioned.
5. Description of the visiting galleries with photographs if possible.
6. Take home messages, conclusion and acknowledgement.

Visit to a Poultry Farm

Poultry farming is applied branch of Zoology which deals with study of rearing of different species of chickens for the commercial production of eggs and meat. Chickens domesticated for eggs are usually called layers while for meat are called broilers. Broilers get marketed within 6 weeks. The poultry is a good cottage industry boostering rural economy of the country. It is the chief source of animal protein in our diet. So visit to the poultry is necessary for the students of Zoology.

Types of Poultry Farm

Two types of poultry farms are available in our country-semi-intensive and intensive poultry farm.

1. Semi-Intensive or Extensive Poultry Farm

In villages or small towns a small poultry farm is established in the courtyard with 80 to 100 poultry birds. The area is separated and fenced with thatched bamboo and covered with net for protection of poultry birds from enemies. Regular supply of poultry feed and pure water is necessary. Usually wooden boxes are provided for the night shelter and protection from adverse situation. The high amount of flesh yielding hybrid varieties like broiler or a good number of egg laying birds, like leghorn, layers, ghagus, *etc.* are reared in these small poultry farms.

2. Intensive or Indoor Poultry Farm

In a square metre 12-15 poultry birds are kept with all facilities for foraging and perching. It is of two types-(i) Battery or cage system, and (ii) Deep Litter system.

In battery or cage system, the cages are placed one upon another; each cage is provided with all amenities to the birds. This the most intensive type of poultry farming. In the deep litter system, poultry birds are kept in large pens up to 250 birds in each floor. The floor remains cover with litters like straw, sawdust or leaves up a depth of 8-12 inches. This system is used for broiler farming.

3. Preparation of Educational Tour Report for Visiting Poultry Farm

Prepare a field report with the following points.

1. Date and time of visiting.
2. Number of co-students with the group, number of teacher guide and number of administrative staff. Take a group photograph.
3. Note the name of poultry farm, location.
4. Note the name of the co-students, name of teacher guide and name of administrative staff. Take a group photograph.
5. Introduction, objective of visiting poultry farm, location, city, district, state, *etc.* and period of visit.
6. Route followed from your place to the farm and mode of journey.
7. Type of poultry farm, housing, intensive or semi-intensive.
8. Name of the poultry breeds, kept in the farm for meat and eggs.
9. The mating of poultry birds, increasing the flocks, improvement in the quality of eggs, meat and other characteristics should be noted.
10. Note the principles of poultry breeding such as the breeding males should be fed with a low protein diet having 13-14 per cent protein and 1-1.5 per cent calcium because high protein diet decreases fertility of male bird. The breeding females should be fed with a high protein diet having 17-18

per cent protein and 3.0-3.5 per cent calcium because protein is required for egg formation (oogenesis). The breeding males and females should be fed vitamin E (40 mg/kg). The breeding males and females should be fed separately in the cage system.

11. In debeaking both males and females are debeaked (cutting of beaks) at 10-14 days of age and again at 12-14 weeks of age. For females leave 2 mm beak from the nostrils. For males cut half of the beak between nostrils and tip of the beak. Provide vitamin K in water 1 day before debeaking and electrolytes for 2 or 3 days after the day of debeaking.
12. Vaccination programme and health care. Fowl cholera vaccine has to be given at about 10 weeks of age.
13. Flock has to be tested for *Mycoplasma* and *Salmonella* at around 16 weeks of age.
14. Deworming should be done every month or once in 6 weeks.
15. LaSota vaccination should be given after deworming.
16. The female breeder should provide rich food with micronutrients in order to obtain high rates of fertility, hatchability and more egg production.
17. Food used for the birds, if possible, composition, *etc.*
18. Man power engaged for production in the farm, and management composition.
19. Information on diseases of poultry birds, if any.
20. The of processing and preservation of eggs by lime water method, sodium silicate method, salicylic acid and cold storage method, wood ashes and salt method.
21. Marketing products and procedure for marketing.
22. Take home message.
23. Conclusion (Ghose and Manna, 2016).

Chapter

18

Laboratory Preparations

Acetocarmine stain: It is chromosomal satin. Take 45 mL glacial acetic acid + 55 mL distilled water and heat to boil in conical flask with cotton plugging. Add 1 gram of acetocarmine powder with stirring. Cool and filter.

Aceto-orcein stain: It is chromosomal satin. Take 50 mL glacial acetic acid + 50 mL distilled water and heat to boil in conical flask with cotton plugging. Add 1 gram of Aceto-orcein powder with stirring. Cool and filter.

Acetoalcohol tissue fixative: Take 25 mL glacial acetic acid + 75 mL ab. alcohol.

Alcohol (70 per cent) for microscopic animals: Take 30 mL dist. water + 70 mL alcohol.

Alizarin red stain: It is used for staining the bones and scales of fishes. It is red stain prepared by dissolving 0.1 g of Alizarin red powder in 100 mL of distilled water. **Acetone-alcohol decolorizer:** Acetone 500 mL +Ethanol or methanol 475 mL + distilled water 25 mL.

Benedict's reagent: For 1000 mL of Benedict's reagent in aqueous preparation add 100 grams of anhydrous sodium carbonate + 1173 grams of sodium citrate + 17.3 grams of copper (II) Sulphate pentahydrate.

Benedict's reagent's purity test: It is blue in color and after heating it remains unchanged.

Benedict's test color observations after test:

Green ppt when sugar is 500-1000 mg/100 mL.

Yellow ppt when sugar is 1000-1500 mg/100 mL.

Orange ppt when sugar is 1500-2000 mg/100 mL.

Red ppt when sugar is 2000 to onwards mg/100 mL.

Boraxcarmine stain: It is skeletal satin. Take 4 per cent aqueous borax solution + 3 per cent carmine solution. Add 70 per cent alcohol with stirring. Cool and filter.

DPX Mountant: It is synthetic resin mounting medium and it dries quickly and preserves stain. It is suitable for hematoxylin-eosin staining.

Eosin stain: It is cytoplasmic stain. Add 1 gram of eosin powder in 100 mL of distilled water for aqueous or 100 mL of 90 per cent alcohol for alcoholic preparation.

Formalin (acetic acid alcohol) tissue fixative: Take 5 mL glacial acetic acid + 85 mL ethyl alcohol (70 per cent) + 10 formalin (40 per cent).

Formalin (neutral) tissue fixative: Take 90 mL tap water + 10 mL formalin + few amount of magnesium carbonate.

Formalin (4 per cent) animal preservation: Take 90 mL tap water + 4 mL formalin. For permanent preservation the solution should be changed after 4-5 days of first preservation. No further change is required.

Fehling's A (5 per cent): 5 grams of $CuSO_4$ in 100 mL dist. Water.

Fehling's B (5 per cent): 5 grams of sodium potassium tartrate in 100 mL distilled water.

Fructose (5 per cent) sample solution: 5 grams of fructose powder in 100 mL dist. water.

Glucose (5 per cent) sample solution: 5 grams of glucose powder in 100 mL dist.water.

Glyceroalcohol: 25 mL of Glycerol + 25 mL of 90 per cent rectified spirit or 90 per cent ethyl alcohol. It is excellent preservative. It is used for preservation of tissues and color of organism. It is good for preservation of specimens.

Iodine (Gram's): 2 g of potassium iodine +300 mL of water+ 1 g iodine crystal.

Iodine (Lugol's): 10 g of potassium iodine +100 mL of water+ 5 g iodine crystal.

Janus Green B: Janus Green B stain is mitochondrial stain. It is prepared by dissolving 0.511 g of Janus Green B dye in 100 mL distilled water.

Labolene cleaning solution: Take 5 mL Labolene and add it to container containing 995 mL of purified water and stir it well. For better results use hot water. Use it for cleaning of laboratory glassware such as bottles, flasks, beakers, and tubing.

Leishman's stain: It is nuclear stain and 0.1 gram of Leishman's stain powder is dissolved in 100 mL of 100 per cent (absolute) methyl alcohol with stirring.

Lipid (5 per cent) sample solution: 5 grams of soaked ground nuts or caster seed or kardai seed or sunflower crushed in 100 mL distilled water and filtered.

Methylene blue stain: It is nuclear stain and 1 gram of methylene blue stain powder is dissolved in 100 mL of absolute ethyl alcohol with stirring.

Molisch's Reagent (10 per cent): 10 grams of α- naphthol in 100 mL ethanol.

Saline Solution, 0.9 per cent Sodium Chloride (NaCl): Dissolve 9 g NaCl (mw 58.44) in 700 mL deionized or distilled water in clean container. Add water to bring total solution volume to 1000 mL.

Schiff's reagent: Solution is available towards the chemical supplier.

Sodium chloride 0.1 M: It is prepared by addition of 5.844 gram of NaCl in 1L and 0.15 M NaCl is prepared by addition of 8.766 gram of NaCl in 1L.

Sodium citrate (0.2 M): Add 29.41 gram of sodium citrate in 1L and 0.2 M sodium citrate is prepared by addition of 5.882 gram of sodium citrate in 1L.

Sodium nitroprusside: 10 grams of sodium nitroprusside redish brown crystals are dissolved in 100 mL dist. water.

Spirit solution: 70 per cent rectified spirit or 90 per cent ethyl alcohol is a very good preservative. It is used for preservation of tissues. Organism initially kept in 50 per cent spirit for a few days and then transferred to 70 per cent.

Starch (1 per cent) solution: Take 1 g of starch powder in 5 mL distilled water and mix thoroughly. Add 95 mL distilled water and heat till boiling point with stirring constantly. Cool and add a few drops of toluene or chloroform as a preservative.

Sudan III: It is fat soluble dye (lysochrome) used for staining protein bound lipid. It included 73.5 mL of 95 per cent ethyl alcohol, warm with water bath add 0.5 gms of Sudan III, stir and add 80 °C distilled water to just below 100 mL mark, stir cool to room temp., dilute to 100 mL distilled water stir then filter.

Sucrose (5 per cent) sample solution: 5 grams of sucrose powder or sugar cane juice in 100 mL distilled water.

Bibliography

Adate, K. J. and Ajagekar, V. V. (2019). In: Zoology. Published by Tejas publication F- 3, Trade Centre, Station Road, Shahupuri, Kolhapur-416 001 and printed at Govind Offset, Y. P. Powarnagar, Kolhapur-416 001: 14-85.

Antram, C. B. (1986). Papilionididae. In: Butterflies of India. Periodical expert book Agency. Vivek Vihar, Delhi: 10-11.

Anschau, A., Caruso, C.S., Kuhn, R.C. and Franco, T.T. (2017). Validation of the sulfo-phospho-vanillin method for the determination of lipid content in Oleaginous microorganisms. *Brazilian Journal of Chemical Engineering*. **34(1):** 19-27.

Atwal, A. S. and Dhaliwal, G. S. (2003). In: Agricultural Pests of South Asia and Their Management. Fourth edition. Published by Mrs. Usha Raj Kumar for Kalyani Publishers, New Delhi. Printed at Xpress Grafics, Delhi:17-286.

Berthold, A.A. (1849). On the transplantation of the testes. *Arch. Anat. Physiol. Wiss. Med., **16**: 42-6.*

Baer, K. E. V. (1827). De ovi mammalium et hominis genesi. *Ovum* of *Mammals.* The outlines the conceptual background of his *discovery* In: Pamphlet.

Bozdogan, H., Erbey, M. and Aksoy, H. A. (2016). Total amount of protein, lipid and carbohydrate of some adult species belong to curculionidae family Coleoptera: Curculionidae). *Journal of Entomology and Zoology Studies.* **4(5)**: 242-248.

Bhaske, S. S. (2018). In: Samanya Vidnyan. Published by Mahesh Shinde's Dnyanadeep Academy for UPSC and MPSC, Sadashiv Peth, Pune-30 and printed at Dnyanadeep Press, Pune: 39-66.

Champion, A. F. (1967). Eexfoliattve vaginal cytology. The role of the general practitioner. *J. College general Practitioner*. **13:** 205-213.

David, B. V. and Ananthakrishnan, T. N. (2006). In: General and Applied Entomology. Tata McGraw- Hill Publishing Company Limited New Delhi. Pritnted at SDR Printers, Loni Road, Shahdara, Delhi: 4-1040.

Deshmukh, S. (2003). In: A Textbook of Histology. Published by A. S. Saini for dominant publishers and distributors for manufacture and export and printed at Orient offset, Delhi-110 053:264-322.

Edwards and Heath (1964). In: Agricultural Pest and their Control (ed.) Deoray, Sonwane and Bedse. Pratibha Prakashan, Satana: 1-6.

Garber, J. C. (2011). Guide for the Care and Use of Laboratory Animals. 11TH Edition. National Academies Press, 500 Fifth Street, NW, Lockbox 285, Washington, Printed in the United St:11-199.

Ghose, K. C. and Manna, B. (2016). In: Practical Zoology. Published and printed at New Central Book Agency (p0 Ltd. Web-offset Division, Dhulagargh, Sankrail, Howrah, Kolkata: 174-297, 339-584.

Gillot, C. (1982). In: Entomolotgy. Published by Plenum publishing corporation, New York. Printed by United States of America: 37-662.

Gomori, G. (1955). Preparation of buffers for use in enzyme studies. *Methods Enzymology.*, **1**:138-146.).

John, J. L. S and Johnson, O. (1931). The determination of uric acid in the study of avian nutrition. *J. Biol. Chem.*, **92**:41-45.

Keele, C. A., and Neil, E. (1964). In: Samson Wright's Applied Physiology. Elventh Edition. E.L.B.S. textbook series, Morrison and Gibb Ltd, London and Edinburg, Oxford medical publications. London: 490.

Leuchs, E. F. (1831). In: Wirkung des speicheles auf Starke. Effect of saliva on starch, Poggendorff's Annalen der Physik und. Chemie. **22:** 623.

Lowry, O. H., Rosebrough, N. J., Farr, A.L., and Randal, J. (1951). Protein measurement with the folin-phenol reagent. *J. Biol. Chem.*, **193**: 265- 275.

Leydig, F. (1850). Zur anatomie der mannlichen Gesch lechtsorgane and Analdrusen der saugethiere. *Z. Wiss, Zool.*, **2**: 1.

McCullagh, D.R. (1932). Dual endocrine activity of testes. *Science N.Y.,* 76: 19

Mutakekar, Y. B., Khabade, S. A., Shine, C. A. and Arabale, V. D. (2013). In: Zoology. Published by Mrs. Bhagyashri M. Phadake, Phadake prakashan, Atharva apartment, Flat NO. 10, Kolhapur-416 012 and printed at Sunil L. Nasirabadakar at Hemant Mudran, H-6, Udyam Co-op, Society, New Industrial Estate Y. P. Powar Nagar, Kolhapur-416 008: 2-71.

Mutakekar, Y. B., Khabade, S. A., Shine, C. A. and Arabale, V. D. (2011). In: Zoology. Published by Mrs. Bhagyashri M. Phadake, Phadake prakashan, Kolhapur-416 012 and printed at Sunil L. Nasirabadakar at Hemant Mudran, H-6, Udyam Co-op, Society, Kolhapur-416 008: 6-261.

Nagbhushanam, R., Kodarkar, M. S. and Sarojini, R. (1996). In: Text Book of Animal Physiology. Published by Mohan Primalani for Oxford and IBH Publishing Co. Pvt. Ltd. 66 Janpath, New Delhi- 110 001 and printed at Chaman Enterprises, 1603, Pataudi House, Daryaganj New Delhi-110 002: 510-564.

Nayar, K. K., Ananthakrishnan, T. N. and David, B. V. (1998). In: General and Applied Entomology. (3rd reprint). Published by Tata McGraw-Hill Publishing Company Limited, New Delhi. Printed at Ram printograph,Delhi: 5-227.

Rastogi, S.C. (1998). In: Essential of Animal Physiology. Published by H. S. Poplai for new age International (P) Ltd. Daryaganj New Delhi-110 002 and printed at Suruchi Printers, North Ghonda Delhi- 110 053:428-502.

Ravindranathan, K. R. (2003). In: Economic Zoology. Published by A. S. Saini for dominant publishers and distributors New Delhi. Printed by Orient Offset, New Delhi.192-194.

Sathe, T. V. and Awate, M.R. (2009). In: Crickets and Household pests. Published by Daya publishing house 1123/74, Deva Ram Park Tri Nagar, Delhi-110 035 and printed in India at Chawla Offset Printers Delhi-10 052: 141-144.

Sertoli, E. (1865). Dell'esistenza di particolari cellule ramificate nei canalicoli seminiferi del testicolo umano. *Morgagni*. **7**:31-40.

Sewak, N. (2016). Biology of spotted bollworm, *Earias vittella* (Fab.) on okra. *Journal of global research computer science and technology*, **4(2)**: 26- 30.

Shinde, C. A., Khabade, S. A., Mutakekar, Y. B. and Arabale, V. D. (2009). In: Zoology. Published by Mrs. Bhagyashri M. Phadake, Phadake prakashan, Atharva apartment, Flat NO. 10, Kolhapur-416 012 and printed at Sunil L. Nasirabadakar at Hemant Mudran, H-6, Udyam Co-op, Society, New Industrial Estate Y.P.Powar Nagar, Kolhapur-416 008: 30-260.

Srinivasan, M. R., Sabarinathan, A., Geetha1, A. Shalini, K. and Sowmiya, M. (2017). A Comparative Study on Staining Techniques for Vaginal Exfoliative Cytology of Rat. *J. of Pharmacol and Clin Res.* 3(3): 555615. DOI: 10.1908

Index

M

N

O

P

Q

R

S

T

U

V

W

Y

Z

9 789359 191263

Printed at Repro India Ltd.